The
Paper Jacket

by Paul A. Matthews

The **Paper Jacket**

by Paul A. Matthews

Wolfe Publishing Company

Manufactured in the United States of America

Wolfe Publishing Company
ISBN: 1-879356-02-3

February 1991

Wolfe Publishing Company
6471 Airpark Drive
Prescott, Arizona 86301

1 2 3 4 5

Some of the old and one of the new! From left to right: (1) a WRA Company .40-90 Sharps Straight, (2) a .44-77 with no headstamp, (3) an 11.43x55R Turkish with no headstamp, (4) one of the author's .45-70s loaded with a 420-grain pure lead hollowpoint bullet, and (5) the old .41 Swiss Vetterli rimfire. These were some of the cartridges that helped build a worldwide reputation for paper patched bullets.

Dedication

Dedicated to that long-forgotten soul who sometime back around 1870, for whatever reason, wrapped a piece of paper around a bullet and upon firing it discovered that it left no leading in the barrel and was perhaps a bit more accurate than the usual undersized bullet or the linen-patched picket bullet. And to my loving wife, Betty, who after 45 years of being married to a nut, still has words of encouragement for that nut to go ahead with whatever project happens to hold the position of top priority.

Warning!

The loads, loading procedures and practices described in this book are those I developed for my own use in my own rifles. Although they proved safe in the rifles in which they were used, they may be dangerous in your rifle. Since I have no control over your loading procedures and practices, nor of the care or quality of your rifles, neither I nor the publisher of this book will assume any responsibility for injury or damage to equipment from the use of any load or loading procedure described herein. If you choose to use the loads and to follow the loading procedures described in this book, you are strictly on your own.

Paul A. Matthews
June 13, 1990

Contents

Introduction

When I wrote *The Practical Paper Patched Bullet*, I did so with a sense of urgency to put on paper that which had been lost to us for almost 90 years. Although there were others who had been and still were using paper patched bullets, very little had ever been written on the subject except for a few paragraphs here and there in articles by Elmer Keith, Frank Mayer and some of the older editions of the Lyman *Ideal Hand Book*. Colonel E. H. Harrison had conducted some in-depth research on hard paper patched bullets for the .30 calibers, and his findings were written up in several fine articles appearing in the *American Rifleman*.

None of the aforementioned writings covered the practical use of the paper patched bullet in today's rifles for today's hunting. Keith, Mayer and the *Ideal Hand Book* were all devoted to black powder as the propellant, while E. H. Harrison wrote of hard bullets which, in my opinion, have limited use in the hunting field by a limited number of hunters.

My concept of a practical paper patched bullet envisions a bullet soft enough to guarantee expansion without breakup, as accurate as the lubricated cast bullet in any given rifle and constructed so as to withstand velocities up to 2,200 fps or a bit faster. As for the propellant, that is the user's choice—black, Pyrodex, smokeless or a duplex load consisting of smokeless as the priming charge and black or Pyrodex for the main charge.

Although I had done a lot of hunting and shooting with paper patched bullets at the time I wrote *The Practical Paper Patched Bullet*, it wasn't until after the book's completion that I settled back and started to delve into the details of this type bullet. It was then that I learned just how much we still didn't know about this projectile, how much had been lost to antiquity and how much erroneous and misleading information had been passed on to us over the years—some of it repeated in my book.

This book is an attempt to correct that, to put on paper the state-of-the-art as it exists today and may well have existed in the late 1800s.

Paul A. Matthews
June 13, 1990

1

Background

The paper patched bullet made its first appearance about 1870 and remained popular as a hunting, target and military projectile until about 1900, give or take a few years. Who first conceived the idea of wrapping a bullet in a piece of paper before firing it down-bore is lost in antiquity, but whoever it was, he or she can rest in peace knowing that they developed the most dependable expanding bullet ever devised by man.

This isn't necessarily my observation, though I fully concur with the finding. It is the remark made by John "Pondoro" Taylor in his fine book *African Rifles and Cartridges*. In speaking of the .450 Black Powder Express, Taylor wrote:

> I killed elephant, rhino and buffalo with the 365-gr. hardened lead bullet, and lion with the soft solid bullet of the same weight from both these rifles.—The solid soft lead bullet is undoubtedly the best and most satisfactory expanding bullet that has ever been devised. It invariably mushrooms perfectly and never breaks up.

Frederick Courteney Selous in his book *Travel and Adventures in South-East Africa* wrote:

> With my .450 bore rifle by Gibbs of Bristol, I have killed ten elephants, using of course the long, solid, hardened

> bullet of 540 grains and a powder charge of only 75 grains.—As I was still very weak (fever) I was afraid the weight of my ten bore rifle would be too much for me, so I decided to see what I could do with my little .450 bore Metford. I used cartridges loaded with 75 grains of powder and long, pointed, toughened 540 grain bullets.

These were paper patched bullets these gentlemen were talking about! And certainly either of these men accumulated more hunting experience in a week's time on the African plains and in the bush than most of us ever hope to see in our entire life. Of special note is the fact that both of these hunters speak of "hardened" or "toughened" bullets, something that to the best of my knowledge was not done in the United States, and for that reason may have given us the erroneous idea that a paper patched bullet *had* to be soft.

Few shooters today recognize the fact that the paper patched bullet was our first controlled-expansion projectile. In fact, most of us associate this feature with metal jacketed bullets and high velocity, and think back to the introduction of the Remington Core-Lokt, the Peters Inner-Belted and the Winchester Silvertip. The facts are, however, that controlled expansion was built into the paper patched bullet by the mere addition of tin to the bullet alloy.

This was made evident in an article appearing in the August 1989 issue of *Magnum* magazine, a publication of South Africa. The author of this article, Walter Lydall, wrote of his grandfather, General Harry Smith Obbard, who served in India during the last half of the 1800s. On Tuesday, February 5, 1878, the general wrote in his diary:

> Bullets for express rifles a twentieth of tin and lead for small game such as buck, etc. a sixteenth for larger game—tigers, etc. Ferris. [Captain Ferris was a squadron commander in the 7th Bangal Cavalry.]

This entry into the general's diary along with other data, implies that in the hunting fields of India reloaded or handloaded ammunition often was used, and that the reloaders themselves were choosy about the hardness of bullet they used for a particular game animal—something we could pay a bit more attention to today.

The British military also was selective on bullet hardness for the .577/450. In an article appearing in *Handloader* o. 145, author Gil Sengel wrote that the British opted for a lead-tin alloy of twelve to one, and that:

> Witnesses said it would shoot through the toughest buffalo hide shield—and the shield's owner—at 500 yards. Given lesser distances it would do the same, only two and sometimes three times over.

This article also states:

> Early military 577/450's were loaded with 85 grains of Curtis & Harveys No. 6 Trade black powder. Bullet weight was 480 grains with a blunt round nose by gunmaker Alexander Henry of Edinburgh, Scotland. Muzzle velocity from the 33-inch barrel registered 1350 fps.

Again, we are speaking of a paper patched bullet, one that was hardened for the express purpose of giving deep penetration through brush, shield and man, and with less deformation than would be apparent with a softer bullet.

This discussion of bullet hardness brings to mind two points or questions. How hard were the bullets that Taylor and Selous used in their .450s while hunting elephant? What were they hardened with? Taylor speaks of using a solid, soft lead bullet on lion, while General Obbard speaks of a hardened bullet, 16 to 1 alloy, for tiger. From my own experience, knowing that the Gould hollowpoint Lyman No. 457122 cast of 1 to 16 tin and lead will expand on deer, I have to believe that what seemed a hard alloy to the general was a soft alloy for Taylor, and that the paper patched bullets used on rhino and elephant were indeed hard, possibly hardened with antimony.

This bring us to another point. Although it was common practice for the African muzzleloader hunter to cast balls for his rifle, after the switch was made to metallic cartridges, very little if any reloading of cartridges in the field was done by the African hunter. At least I have yet to read about it. This is not surprising as it was much simpler to carry factory-loaded ammunition on an extended elephant hunt than to carry the powder, primers, cases, moulds, bullet metal and reloading tools.

Not so in India where most hunting trips were made by military officers and were of short duration and strictly for recreation instead of profit. Here, loading your own ammunition was popular and added to the thrill of hunting. Going back to the article from the South African magazine *Magnum*, author Walter Lydall tells of finding a clipping of a question and answer column taken from the Indian newspaper *Pioneer*. Further research turned up the following question in a column dated January 9, 1878.

> Express Rifles—I shall feel extremely obliged to sportsmen for information on the following points: 1. What is the best substitute for the Russian tallow lubricators for .450 express rifles? 2. Is it necessary to use only the prepared paper as a covering for the bullet, or will not any ordinary kind of paper answer as well? 3. What is the quantity of powder required for .450 express bullets for a Rigby rifle? 4. Does it not conduce to safety and speedier explosions to shake the powder in the cartridge in order to allow the grain to settle down compactly before putting in the bullet?—Shikaree.

On January 19, 1878, two different gentlemen signing themselves Aaester and Nahar answered Shikaree's question as follows:

> Express Rifles—In reply to Shikaree's queries in the *Pioneer* of the 9th instant—(1) The best to use are Eley's wax lubricators, sold in Bombay in boxes of 100 each, at 8 annas per box, which also includes 200 thin jute wads, one to be placed on either side of the wax wad. (2) Prepared paper only should be used for covering the bullets; common paper would be apt to strip, and cause leading and inaccuracy. Granted a faultless bore, the secret of accuracy lies in preparing the bullet properly. (3) The quantity of powder required depends upon the capacity of the cartridge case, which varies from 3¾'s to 6 drams. If

Shikaree uses solid-drawn cases, which, with improved machinery, have now an almost uniform capacity, and if his rifle is properly chambered so as to admit of as little expansion in the cartridge case as possible when fired, he can always use each cartridge case as its own measure, it being merely necessary to fill them up until there is a space left at the top equal to the thickness of the three wads and one-third the length of the bullet. A plug of wood, the diameter of the mouth of the case, and marked round with a line of ink, will guide him in measuring this space correctly. (4) The cases may be tapped a few times on a table as the quickest way of loading by hand, but the best plan is to pour the powder in slowly through a funnel with a contracted mouth; it is not generally known, but it makes all the difference, as regards the amount of powder a cartridge case will hold, whether the powder is poured in very slowly or quickly.

N. B.—No matter how carefully a cartridge may be loaded, after being in the pocket for several days, the bullet will require more or less pushing down again; an empty space behind the bullet spoils accuracy, and causes a rifle to shoot low.—Aaester.

Ditto—(1) The best substitute for Russian tallow is bees' wax and hogs' lard in equal parts. The cavities of hollow expanding bullets should be filled in with bees' wax only. For heavy game the bullets are fitted with copper tubes, which prevent their breaking up too easily; for black-buck shooting, these bullets are not required, as bullets filled with bees' wax are quite sufficient. (2) It is necessary to use only the prepared paper for wrapping bullets, for the following reason:

Expansive bullets require to be most accurately fitted to the bore, and the prepared paper, when wrapped over the bullets and lubricated, fulfills this object. Other paper, if used, may be too thick or too thin, and not sufficiently tough for the purpose. (3) The ordinary charge for an Express .450 bore is 3½ drams or 96 grains; some Express rifles take 4 drams or 110 grains. All expansive bullets have one charge of one kind of powder, with which they are found to act best. Great irregularity of shooting takes place on slightly increasing or diminishing such charges.

> Uniform strength of powder is a matter of great importance. I find Curtis and Harvey's No. 6 powder very good. (4) The object of granulating powder is to enable the flame to play freely amongst the mass. This is nullified if the charge is tightly rammed down. It is not advisable to shake the powder in the cartridge; but it is essential that there shall be no space between the lubricating wad and the powder, and between the wad and the bullet.—Nahar.

Reading these questions and answers, it is easy to see that many of the same questions we ask today were asked well over 100 years ago. The one surprising thing to me is relative to the comments about shaking the powder down in the cartridge case. Apparently none of the shooters involved were aware of the fact that using a drop-tube when loading black powder settles the powder in the case to its maximum without crushing or fracturing the kernels.

During the last three decades of the nineteenth century, especially in the 1870s and 1880s, the paper patched bullet ranked supreme above all others. It was widely used on every hunting field in the world on all types of game. Most major armies of the world, except the U.S., depended on the paper patched bullet come rain or shine, and it was a widely used favorite on the target range being used in breech loading cartridge rifles, percussion muzzleloaders and the muzzleloading breech loader.

To grasp some idea of the popularity of the paper patched bullet, one has only to look at some of the old catalogs reprinted in various issues of *Gun Digest*. In this publication's sixth edition in the article "Winchester's Forgotten Cartridges, 1886—1900," there are 19 cartridges illustrated which are loaded with a paper patched bullet. These range from the old .30-30-165 Frank Wesson up through the .45-120-550 cartridge for the 3¼ Sharps.

Again, in the seventh edition of *Gun Digest* we are shown an illustration of an 1888 UMC cartridge display board on which 35 cartridges are loaded with paper patched bullets. And again in the ninth edition of the same book, we find a reproduction of an 1884 Kynoch catalog with life-size illustrations of six military cartridges, three express cartridges, 10 cartridges for Westley Richards rifles and four cartridges for Gatling guns—all with paper patched bullets.

Finally, in the 11th edition of *Gun Digest* in an article by Graham Burnside entitled "Sharps Rifle Company Ammunition," there are 40 different cartridges illustrated with paper patched bullets. So, for a while at least, the paper patched bullet ranked supreme in many armies of the world, as well as in the greatest game fields the world has ever seen. Its passing was a regrettable loss.

For all of its popularity and widespread use, however, the paper patched bullet never reached a peak of development in the United States comparable to that of England and Europe. I believe there are three reasons for this:

First, the U.S. military opted for the grooved, lubricated bullet when the .45-70 cartridge was adopted in 1873.

Secondly, the game to be hunted on the North American continent did not impose the penetration requirements of that found in Africa. Although the buffalo was massive in structure and had a thick tough hide, the animal seldom fought back and bullet weight in itself was sufficient to guarantee the necessary penetration for positive kills.

Thirdly, being a highly competitive society, Americans were then and are today always willing to compromise a desirable feature in favor of productivity. It is simply a lot easier and faster to produce grooved lubricated bullets than it is to hand-wrap a piece of paper on each individual bullet as it comes from the mould or swaging die. If the grooved lubricated bullets had a tendency to lead the barrel, then it was up to the consumer to clean it out more often.

To the best of my knowledge paper patches were always applied by hand; there was never any machine developed to do the job. Girls and women always did the patching because, it was said, their fingers were more nimble and their saliva, used to wet the patch, adhered to the bullet better than a man's. Phil Sharpe, on page 104 of his book *Complete Guide to Handloading* states:

> At the ammunition plants girls rolled these [patches] on, and many girls were capable of patching 12,000 bullets each per working day.

That is a lot of bullets in one working day! And if you ever look at some of the factory-produced paper patched bullets, you'll find every one of them done to perfection and identical to the one before.

Most of us, when we think of paper patched bullets, envision finger-length cartridges made to fit a bore large enough to swallow your little finger down to the first knuckle. That, however, was not the true situation. Although the majority of paper patched bullets were for .30-caliber and up, even the smaller calibers often carried the paper jacket. On page 43 of Ned Roberts' and Ken Waters' fine book *The Breech-Loading Single-Shot Rifle*, there appears the following:

> During the years the Maynard rifle was in production, it was furnished in many different calibers, several of which were special or peculiar to that rifle only. Among these were the .22-10 caliber Maynard centerfire cartridge using a 45-grain paper patched bullet, undoubtedly the smallest paper patched bullet ever in regular production . . .

Make no mistake about it, trying to wrap a paper patch on a .22-caliber bullet has got to be the supreme test for nimble fingers—certainly not for mine which do well to handle .30-caliber bullets and upward to the .54 Minie ball. The very fact, however, that such small paper patched bullets were in production is solid evidence of their popularity for the short span of three decades.

Why then, if the paper patched bullet was so popular, if it was the ultimate expanding bullet in the hunting field and if it was the one bullet that could have cast-in controlled expansion, why was it so short-lived?

Two reasons: metal patched bullets and economics. Metal patched or jacketed bullets came on the market during the same time frame as smokeless powder, and could be punched out by machinery at a lot less labor cost than the hand-wrapped paper patched bullet. As for the shooter who continued to make his own, it was simpler and easier to cast a grooved lubricated bullet than to take the time and patience to apply paper patches. It mattered not that a paper patched bullet does not lead the barrel—indeed, I have *removed* lead from a barrel with paper patched bullets—or that for the 2,000 fps velocities of that day paper patched bullets would work as well as the new-fangled metal jacketed bullets, the paper patch had to go. Too much hand labor involved!

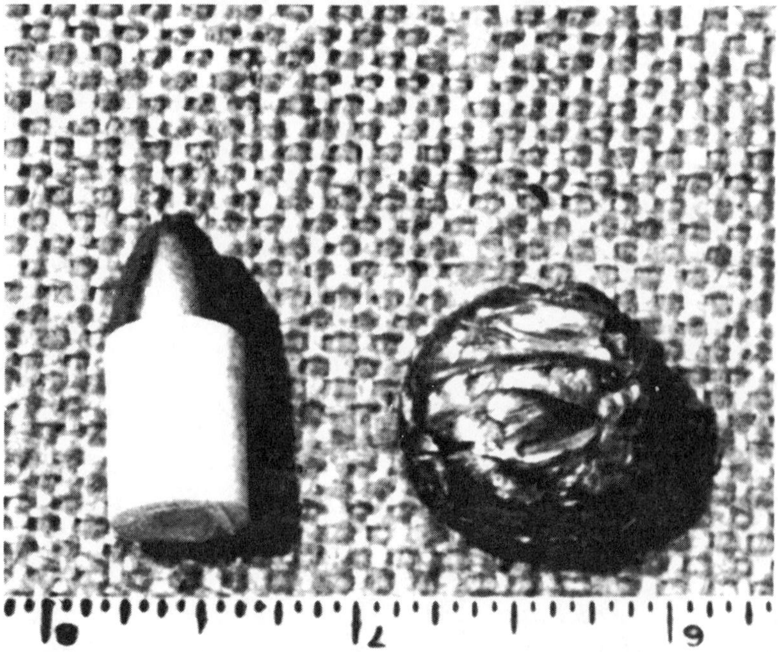

This is the kind of expansion John "Pondoro" Taylor was so enthusiastic about. On the left is an unfired 385-grain .45-70 paper patched bullet cast of pure lead with 3 percent tin. On the right is an identical bullet from the same melt retrieved from a buck after traveling through 18 inches of muscle and bone. The bullet mushroomed out to over twice its original size, 15/16 inch, and lost only nine grains weight—about 2 percent. Load was 53 grains of IMR-3031.

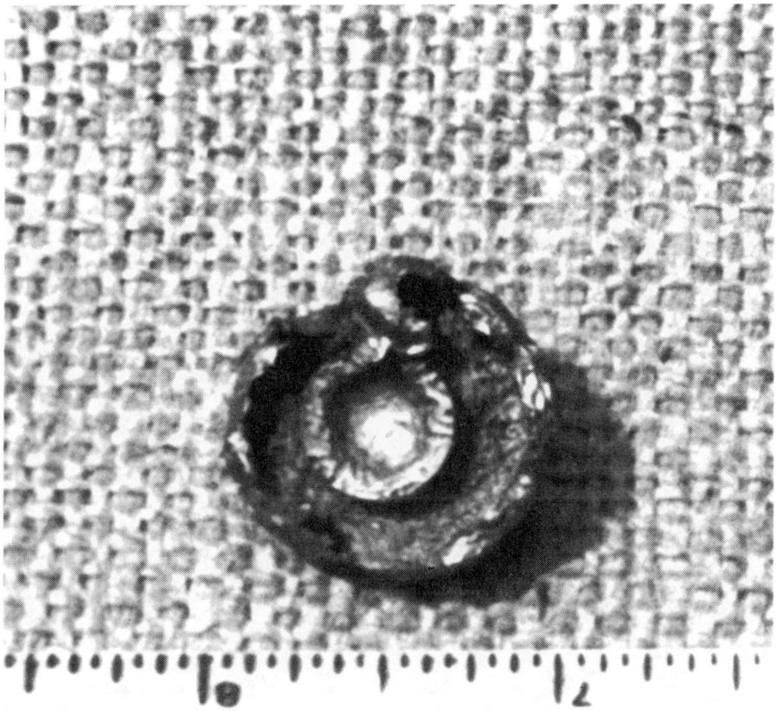

This is the base of the bullet taken from the aforementioned buck. Note the indentations around the base's rim made by the powder kernels at the time of ignition.

This 200-pound wild hog took a single 405-grain soft paper patched bullet through the chest at about 50 yards. He dropped on the spot. Load was 53 grains of IMR-3031.

This is the Lyman Minie ball mould No. 533476 with two bottom plugs, one to form the conventional cone-shaped base cavity, and the other to form a cupped base. The Minie balls are patched with a .003-inch-thick paper to provide a snug fit in the bore of a .54-caliber flintlock. The small circular tin contains Rooster Laboratories' Minie Lube for lubricating the patch.

Close-up view of naked and patched Minie balls. Note that the base of one Minie ball is filled with lubricant.

2

The Bullet

My earliest acquaintance with the paper patched bullet dates back about 55 years to a time before I even knew what it was I was looking at. An old barn about half a mile down the road from our house was being renovated, and tucked away on top of one of the beams above a horse stall was a small, square cardboard box of commercial paper patched bullets. I have no idea what caliber they were, at least .38 and maybe as large as .45. But I do remember the excitement that surged through my stomach when Tom Harris, a lifelong friend, handed me one of those heavy slugs and I bounced it in the palm of my hand. Would that I'd had the sense then to study that box of bullets and possibly trade a dozen marbles or jack-knife for the works—that is, assuming I had the marbles or jackknife.

The next time I brushed shoulders with a paper patched bullet was one rainy day in the early 1950s when Harry Stebbins showed up at the house with an original Sharps side-hammer chambered for the .44-77 and several boxes of factory-loaded cartridges sporting soft lead, paper patched bullets. Again, neither of us fully grasped the significance of what we held in our hands and we proceeded that afternoon to fire every single round of ammunition from the front seat of Harry's station wagon at stones embedded in the far bank of Mallory Run. Somewhere in that experience there should have been at least a shred of intelligence that said to save at least one box of those cartridges, but if such a shred existed, it was buried under a cloud of black powder smoke and flying fragments of stone.

During the summer of 1953 while doing a lot of shooting with the .375 H&H Magnum, I began thinking about paper patched bullets. And when the Lyman *Ideal Handbook No. 39* came out in early June or July of that year, there on page 182 was a cutaway drawing of the old Ideal Cylindrical Adjustable Mould for casting paper patched bullets. At the bottom of the page was the following note:

> Bullets cast in the cylindrical mould are so much better that we will *not make any more* hinged moulds for patched bullets. Send stamps for sample bullet with either round or flat point.

The Ideal Cylindrical Mould was advertised in .25, .32, .38, .40 and .45 calibers. It was an adjustable mould capable of casting the full range of bullet weights for each caliber listed, and the bullets were easily ejected from the mould by pressing against the nose-forming punch. This, of course, is assuming that a lead or lead-tin alloy having a high shrinkage factor was used. An alloy rich in antimony might not eject so easily.

The bullets cast in the Ideal Cylindrical Mould had parallel sides and a perfectly flat base. That is, there was no cup or shallow cavity in the base in which to tuck the tail end of the patch. Much was said about the parallel sides. In their description of the mould, the Ideal company said:

> . . . with it the bullets are punched out of the mould leaving them perfectly true *without taper*, assuring the whole length of the bullet bearing in the barrel of the rifle, therefore not requiring so much upsettage and obviating any possible chance of it being started in a tipping or crosswise manner, so common with the tapered bullet.

Although this may have a taste of Madison Avenue advertising fluff, it implies that paper patched bullets cast in a split-block mould had a slight taper from the base to the ogive's origin. In fact, taken in conjunction with their earlier statement, we know this to be a fact with the Ideal split-block moulds. This is a bit surprising, however, because rolling a patch on a tapered bullet is not the easiest thing to do as the bullet doesn't roll in a straight line, thus making it difficult to keep the leading edge of the patch square with the bullet's axis.

Be that as it may, the Ideal Cylindrical Mould was an excellent mould and will produce superb bullets. Shortly after seeing the cutaway drawing in the Lyman handbook in 1953, I made such a mould for the .375 H&H, and for the first time in my life cast, patched, loaded and fired a paper patched bullet. While I am by no stretch of the imagination an accomplished machinist, as evidenced by the fact that my mould cavity had about 16 sides to it, those early experiments with paper patched bullets showed a lot of promise.

One thing that I didn't care for about the Ideal Cylindrical Mould was that it leaves an abrupt shoulder, however slight, at the ogive's origin. I like to see a smooth, unbroken contour here, both for aesthetic and patching purposes. If the leading edge of the patch is carried 1/32 inch beyond the origin of the ogive on a smoothly contoured bullet, it will shrink tightly to the bullet and will enter the rifling with much less chance of stripping or tearing.

My own preference in a paper patched bullet is one that follows traditional lines and has a cupped base into which I can turn the overhanging edge of the patch. For a rifle having a freebored chamber, I much prefer a bullet having parallel sides all the way to the origin of the ogive. For a rifle that does not have the freebore, however, I want a bullet having a bit more than a full caliber of parallel sides followed by a slight taper to the ogive's origin.

This parallel side business is extremely important if you are to be able to use the full capacity of the cartridge case.

With a freebored chamber on a single-shot rifle, the bullet having long parallel sides can be seated well out to just shy of the rifling to deliver good accuracy and maximum case capacity.

If the same bullet is used in a rifle that has no freebore, however, then it has to be seated so deeply in the case so as not to interfere with the rifling, that case capacity is compromised. But if the bullet used in a rifle without the freebore has parallel sides approximately $\frac{1}{8}$ inch longer than a full caliber, and then has a slight taper to the ogive's origin, it can still be seated out without interference with the rifling or compromising case capacity.

I ran into this problem head-on when my brother Allan and I were preparing for a boar hunt in North Carolina. Each of us was using a Ruger No. 3 .45-70 with paper patched bullets. My rifle had the freebore, his did not.

For my rifle I used 405-grain bullets cast of lead with 3 percent tin in a Hoch bullet mould made by Dave Farmer of Colorado Shooter's Supply. This was a Hoch design, not my own, having parallel sides with a short radius ogive terminating in a $\frac{3}{16}$-inch diameter meplat. The base of the bullet has a shallow cupped recess with a $\frac{1}{16}$-inch wide rim. It is a superb bullet, casting .452 inch in diameter when cast of my soft alloy. I patched this bullet with a one-inch wide patch covering $\frac{23}{32}$ inch of the bullet shank.

With my freebored chamber I could load this bullet on top of 53 grains of IMR-3031, leaving $\frac{9}{32}$ inch of exposed patch so that it just kissed the rifling, and still chamber the cartridge with no perceptible interference. For Allan's rifle which did not have the freebore, however, I had to use a shorter 389-grain bullet from the same mould and reduce the powder charge to 51 grains of IMR-3031.

Now although it might be rightfully argued that the boar Allan shot didn't know the difference, there is no argument to the fact that Allan's load was compromised due to the bullet's shape.

Since that trip to North Carolina in 1988, Fred Cornell made me a bullet swaging press and a set of dies for swaging paper patched bullets. Bullets from this die have a long radius ogive, a $\frac{9}{32}$-inch diameter meplat and $\frac{1}{2}$-inch parallel sides on a bullet weighing approximately 415 grains. The patch covers $\frac{19}{32}$ inch of the shank as opposed to $\frac{23}{32}$ inch for the Hoch bullet. With this new bullet in Allan's rifle, I can load a full 53 grains of IMR-3031 and not run into any interference when chambering a cartridge.

Loading paper patched bullets for the Marlin 1895SS lever action rifle where the overall length of the cartridge is restricted to 2.55 inches brings into focus yet another requirement—and don't think for a minute that these rifles won't handle soft paper patched bullets—they will! Because of the overall length limitation, however, practical bullet weight is restricted to no more than 420 grains. Paper patched bullets for the Marlin 1895SS should have parallel sides with a very short ogive and a large diameter meplat. In short, put all the weight possible in as short a package as possible. Then regulate the width of the patch so that when the bullet is seated to its proper overall length, the leading edge of the patch is flush or slightly below flush with the case mouth. This prevents mutilation of the patch when stuffing cartridges through the loading gate into the magazine.

Although I much prefer the smooth-sided paper patched bullet, there is another style on the scene that is quite popular and has certain advantages. This bullet has a series of very shallow grooves the length of the bullet shank and a shoulder at the origin of the ogive. However, on this bullet the origin of the ogive is larger in diameter than the shank and provides a positive locating point for the patch's leading edge. This not only assures that the leading edge of the patch is the same on every bullet, but also offers protection to the patch as the bullet engages the rifling.

Although the shallow grooves along the bullet's shank are supposed to provide a better grip for the patch, I believe they serve a far better function of holding patch lubricant. Any patch that is properly applied grips the bullet tightly enough so that it cannot be twisted or turned by hand, grooves or not. These bullets also have a flat base, and the moulds made for them are not adjustable for bullet weight.

Gunscribe Mike Nesbitt has done a lot of work with this type bullet. He has used them in various calibers, primarily with black powder or duplex loads, and has taken a number of game animals with them. Although this style bullet gives superb performance at black powder velocities, there has been some criticism that, because of the groove diameter ogive, the bullet has a tendency to lead the barrel at higher velocities. I can neither confirm nor deny this, although I suspect it is an accurate statement. In my own loading, I make certain that no exposed portion of the bullet contacts the barrel.

Another bullet which serves well with a paper jacket, but is sadly overlooked, is the common garden variety cast bullet with lubricating grooves. For the person who wants to stick his toe in the water just to see what this paper patched bullet business is all about, the conventional cast bullet will work just fine. I like to run them through a sizer-lubricator die before patching to get them as small as possible, and then run them through the same die again after patching and lubing to get them down to the proper size. We will discuss this in detail later on.

Now that we have discussed the shape and design of paper patched bullets, we should examine bullet diameter as it relates to the barrel and the propellant being used. Just as the bullet's shape must be matched with the chamber, so also must its diameter be matched to the rifle's bore *and* the propellant by which that bullet is expelled from the barrel.

Ninety-nine percent of the success with a paper patched bullet propelled by smokeless powder is directly related to the diameter of the unpatched bullet; it doesn't matter whether we are dealing with hard bullets or soft. Make no mistake—smokeless powder does not bump up the diameter of a bullet, hard or soft, to any appreciable degree. Years ago I would have argued against that statement; but today I know better.

Anyone contemplating the purchase of a mould for paper patched bullets should first slug the barrel of their rifle and determine the bore diameter—across the tops of the lands—as well as the groove diameter. Although groove diameter is the most important of the two measurements for the grooved lubricated bullet, it is the bore diameter that dictates the desired size of the paper patched bullet.

Measuring bore diameter is not always an easy thing to do, certainly not as simple as measuring the groove diameter where all you must do is push the slug through the barrel and measure across it with a micrometer. One of the best ways I have found to measure bore diameter is to cast a number of bullets of the caliber involved from a mould having a fairly long parallel-sided nose. Then, by squeezing the bullet endwise very slightly in a vise or on the loading press, increase the diameter of the nose until it will just enter the bore leaving only the slightest trace of rifling marks. The more indistinct the trace is, the more accurate will be the measurement. Once the nose is squeezed to the proper size, measure its diameter with a micrometer.

Paper patched bullets for use with smokeless powder should have an unpatched diameter equal to or .0005 inch over the rifle's bore diameter. If you order a custom mould to produce a bullet of this dimension, have the dimension based on pure lead so that any bullet cast of any alloy from that particular mould will deliver good performance in the rifle for which it was ordered. In other words, if the bullet is ordered to measure .450 inch in diameter when cast of pure lead, it will measure approximately .4515 inch when cast of Linotype and will give superb performance.

Conversely, if the same mould is ordered to measure .450 inch when cast of Linotype, bullets cast of pure lead or a lead-tin alloy will be a shade too small for use with smokeless powder.

When Charlie Canoll made my first .45-caliber mould, he asked me what size I wanted and I said .450 with pure lead. That was exactly what I got for use in my Ruger rifles, both of which had a bore diameter of .4515 inch. Fortunately, the gods were smiling and the first bullets I cast were of pure lead and weighed 355 grains. These were loaded in front of 55 grains of IMR-3031 for a velocity of 1,945 fps and shot into a group at 100 yards that measured 1 15/16 inches horizontally by 3/8 inch vertically. Time after time this load shot into groups as small as 1 1/4 inches, and never over two inches.

The fly in the ointment came about when I tried to vary the load either by going to a reduced powder charge of a faster-burning powder or going to a heavier bullet with a lighter charge of IMR-3031. Even though the bullets remained end-on, they did not group well on the target. Then one day in a fit of desperation, I tried a very heavy bullet, about 500 grains, hardened with about 3 percent tin and backed with a hefty load of IMR-3031. Upon firing, the patch was stripped from the bullet in the barrel, the barrel was badly leaded and the bullet keyholed at 50 yards.

In analyzing the sequence of events here, it appeared that pressures with 55 grains of IMR-3031 were just enough to slug the 355-grain bullet in the barrel and thus give good accuracy. Any reduction in powder charge, even with heavier bullets, failed to do this sufficiently enough for good accuracy, and when the bullets were hardened to any degree, there was no bullet upset whatsoever.

To prove my analysis, I had Harold Reynolds, a top flight engineer and superb craftsman, make me a "bumping-up" die so that I could increase the diameter of my unpatched bullets to .452 inch. As soon as I did this, my problems vanished and I found that I could use paper patched bullets of any hardness and with any load as long as the unpatched bullet's diameter was equal to or .0005 inch larger than the barrel's bore diameter.

Bullets to be used with black powder or Pyrodex bring into focus an entirely new set of parameters. Now we are faced with the problem of interference with powder fouling in the barrel's throat. In short, you can't chamber a paper patched bullet that measures .459 inch across the patch in a barrel throat that measures about .455 inch when fouled with black powder crud. It's like trying to put 10 pounds of sugar in a five pound bag—it won't fit!

The diameter of the *unpatched* bullet for use with black powder or Pyrodex should be approximately .004 to .006 inch smaller than the rifle's bore diameter, so that after it is patched it is .001 or .002 inch larger than the bore diameter. Frank Mayer, one of the last of the buffalo hunters to depart this earth, put it this way in the December 1934 issue of the *American Rifleman*:

> Get a mould that will cast bullets of such diameter that when patched they will be just a trifle too snug a fit to enter the muzzle of the rifle, but will still permit fairly easy entrance of the cartridge in the rifle chamber.

All of this, of course, is assuming that the barrel will not be cleaned between shots as is usually the case while hunting. If a barrel is cleaned after each shot, then the same size bullet can be used for black powder or Pyrodex as is used with smokeless loads. The same also applies to duplex loads, provided the smokeless priming charge is sufficient to burn off most of the solids.

This business of bullet diameter and shape is something that the average shooter usually doesn't consider until it is too late. A few years ago I purchased a Ruger No. 1 chambered for the .458 Winchester Magnum cartridge. The chamber on this rifle is cut in accordance with SAAMI standards and is just about as perfect a production chamber as there is for cast bullets, whether they are grooved lubricated or paper patched. The .458's throat has a taper of slightly less than half a degree beginning at the end of the chamber and extending up the barrel for a bit better than one inch.

I bought this rifle with the idea of using it with black powder, and it is superbly accurate when used in that manner. But when I first loaded a handful of cartridges with 90 grains of FFFg behind a 500-grain plus paper patched bullet and hustled to the range, I failed to take into account the problem of black powder fouling. Obviously, with a clean barrel the first cartridge chambered with no problem, but the second cartridge wouldn't chamber within ¼ inch! (This was with bullets having an unpatched diameter of .450 inch.)

Later I used a duplex load in this rifle consisting of 72 grains of FFg and 12 grains of SR-4759 with a 565-grain paper patched bullet having an unpatched diameter of .450 inch. This load was quite accurate and gave no chambering problems at all.

There is one more possible situation affecting bullet diameter which might concern a user of paper patched bullets. This event occurs when loading the paper patched bullet from the muzzle. I have done this in a Navy Arms .45-70 rolling block, and Navy Arms also offers a muzzleloading percussion rifle designed especially for this type of loading. In a situation such as this, we are definitely using black powder or Pyrodex along with a very soft bullet, quite often pure lead.

Paper patched bullets used for muzzleloading should be a snug, thumb-push fit in the rifle's muzzle. This means that the unpatched diameter of the bullet should be approximately .009 inch smaller than the bore diameter of the rifle, and the thickness of the patch regulated to provide the proper fit.

Occasionally muzzleloading paper patched bullets can greatly increase a rifle's versatility. For example, you will have a tough time getting more than 65 grains of black powder in a .45-70 case and still have room to seat a bullet.

However, by placing a primed case in the chamber and *blocking the hammer so there is no possibility of it falling against the firing pin*, you can dump 100 or 110 grains of FFg down the muzzle, seat a 550-grain paper patched bullet and duplicate the old .45-100-550. When I did this with my Navy Arms rolling block, I slipped a 1⅛ inch diameter steel ring made of .176 inch stock over the hammer's nose and then let the hammer down against the breechblock. The ring was then held captive against the breechblock, and the hammer couldn't contact the firing pin.

I also used paper patched Minie balls in my .54-caliber flintlock. These were cast of pure lead in a Lyman No. 533476 mould and patched up to .542 inch in diameter. This made a thumb-push fit in the barrel. They were pure devastation on deer.

Fred Cornell made a special .45-caliber muzzleloading rifle with a heavy 1⅛-inch octagon barrel, 32 inches long with a 20-inch twist for shooting heavy paper patched bullets. His load was 90 grains of FFFg behind a 550-grain bullet, and his target was a red thumbtack in the center of a white sheet of paper at 100 yards!

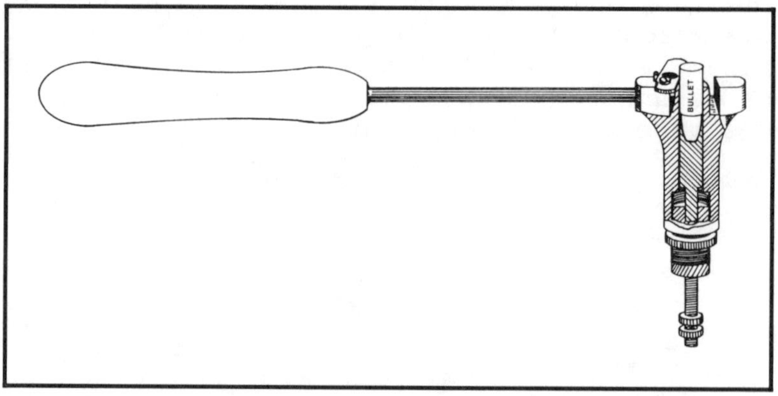

The Ideal Cylindrical Adjustable Mould is a superb practical mould for producing paper patched bullets and is easily made on a lathe by any skilled machinist. With one of these moulds the shooter can cast a full range of bullet weights for any given caliber. These moulds originally were offered by the Ideal Manufacturing Company in .25, .32, .38, .40 and .45 calibers. As shown here, the sprue plate has been swung to one side and the bullet partially ejected by pushing upward on the nose-forming plunger. Length of the bullet is determined by location of the threaded bushing in the mould's bottom.

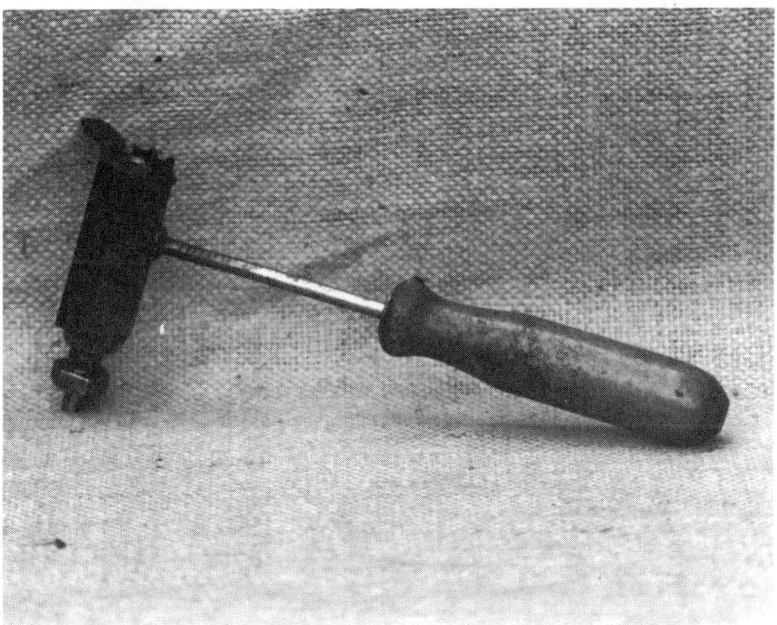

A cylindrical adjustable mould made by the author in 1953 for producing paper patched bullets for the .375 H&H Magnum. This was the author's first mould for casting paper patched bullets.

This five-shot group was fired with pure-lead paper patched bullets at a chronographed velocity of 1,945 fps. The bullets had an *unpatched* diameter of .450 inch, and were patched with 16-pound 25 percent cotton fiber paper.

Fred Cornell at the shooting bench with his .45-caliber slug gun shooting at thumbtacks at 100 yards!

3

Casting the Paper Patched Bullet

There has been so much written in recent years about bullet casting that it seems almost superfluous to discuss it here. Yet in reading much of what has been written, I find that most of it pertains to grooved bullets cast of lead-tin-antimony alloys not too dissimilar from Linotype metal. Casting with these alloys in .30-caliber rifle moulds and the usual assortment of pistol moulds is far different from casting pure lead or a very soft alloy in a mould that gobbles 400 grains or more at a single gulp. Many times I have read about the problems involved in casting soft lead roundballs and Minie balls for the muzzleloaders, when the basic problem was that the caster had neglected the much higher melting temperature of lead.

For example, it is common practice to set the thermostat anywhere from 700 to 750 degrees Fahrenheit for Linotype metal, which starts to melt at about 460 degrees. This temperature gives us a very fluid melt which nicely fills out the mould.

Pure lead, on the other hand, melts at 621 degrees and requires a much higher temperature to get the fluidity necessary to fill out large moulds and to keep the lead molten until the entire bullet is poured. With an electric pot, 850 degrees is not overly hot for pure lead. In fact, you may have problems with fill-out (the metal cooling too fast), especially if the thermostat has some age to it and doesn't really bring the pot up to 850 degrees. Under these conditions, the addition of a small amount of tin helps to lower the temperature required and simultaneously assists in filling out the mould.

For me, the proper temperature is a matter of judgment rather than instrumentation. When the surface of the melt takes on a light straw color, it's hot enough. If you let it get hotter, the surface turns a crinkled blue and starts to oxidize, but with the straw color it flows freely with little oxidation. It usually takes a dozen or so bullets before the mould is hot enough to provide a perfect base, but once the mould is properly heated, almost every bullet comes out perfect.

As with any other type of bullet, perfection is the name of the game. Don't waste your time trying to salvage bullets with wrinkles or a rounded edge base. Get the metal hot, pour a generous sprue and accept nothing but the best. Let the bullets fall onto a thick padding to prevent nicks, bruises and distortion.

One other thing I learned years ago, was to avoid striking the sprue plate with a stick or plastic mallet. Why pay good money for a fine precision instrument and then proceed to beat it to death with a club? Get a pair of good gloves and work the sprue plate with your hand. When the mould is cold, it *might* be necessary to sometimes use added persuasion on the sprue plate, but after the mould starts to heat, you can easily swing the sprue plate by hand.

Casting good paper patched bullets starts with quality equipment—moulds, furnace and bullet metal.

It is difficult to beat a nose-pour bullet mould regardless of whether you are casting the grooved or smooth-sided variety. When casting bullets, the major problem is voids—small bubbles of air usually trapped in the bullet just under the sprue. If these voids are on the bullet's axis, little damage is done other than a variation of bullet weight. But if the void is a bit irregular or off center because the sprue plate slipped during the casting process, or was deliberately put off center, then the bullet is off-balanced. This condition is far more likely to occur and is far more disastrous to accuracy with a base-pour mould than with a nose-pour mould.

There are many producers of fine bullet moulds, both production and custom. However, it is my belief that if a shooter is really serious about paper patched bullets, his best move is to order an adjustable, cupped-base, nose-pour mould from a custom maker. I originally ordered two custom nose-pour moulds for grooved lubricated bullets from Richard Hoch, and both of these proved superb. After he sold his mould business to Dave Farmer of Colorado Shooter's Supply, I ordered an adjustable, cupped-base, nose-pour mould for casting smooth-sided .45-caliber paper patched bullets weighing from 350 to 550 grains. It is an excellent mould and I have cast thousands of bullets with it.

When it comes to the casting process, it makes little difference whether you favor the dipper method or the bottom-pour, gravity-fed electric melting furnace. I've gone both routes and produced good bullets by each method. I started out in 1949 with a dipper and a gasoline plumber's furnace, and about 1962 I switched to a Lyman Mould Master electric furnace with its bottom-pour spout. I used the bottom-pour valve until about 1980 and then one evening I took the valve off and plugged the hole with a ball bearing. I've used the dipper ever since and will continue to do so as long as I cast bullets. I get better bullets with a dipper than with the bottom-pour gravity feed.

One reason I get better bullets by dipping is because I don't try to see how many bullets I can cast in an hour, and I keep my metal hot—very hot. In fact, when the thermostat on my Lyman Mould Master started to deteriorate for the second time, I merely ran the wires past the thermostat so that the pot heats at full throttle all the time. Even so, this will not keep pure lead hot enough to cast large, heavy bullets without traces of wrinkles, so today

I use my Mould Master only when I'm casting from a mixture of God knows what—wheelweights, backstop scrap or whatever—metal that melts and pours at a considerably lower temperature than pure lead.

For casting my hunting bullets—and these are either pure lead or lead with 3 percent tin—I use a small Coleman Model 508 camp stove. This little stove will melt pure lead with heat to spare and can be regulated easily. With it, I use three different melting pots, one containing pure lead, one with lead and 3 percent tin, and the third with type metal. The Coleman stove melts the metal in a fraction of the time required by the Lyman Mould Master, and the tank of fuel will last an hour and a half to two hours, depending upon its regulation. I seldom cast longer than an hour and a half at a time, so for this purpose the Coleman stove provides an excellent heat source.

When it comes to converting lead pipe, cable sheathing, wheelweights, etc., to the more manageable-to-handle ingots, you can do little better than a 20-pound tank of liquid propane with a plumber's burner on top and a cast iron pot that will hold 20 to 30 pounds of melt. This and a large cast iron ladle and about three of the Lyman ingot moulds will let you convert 100 pounds of scrap into clean, ready-to-use ingots in about two hours. Flux the metal well with paraffin or something similar to get the dirt out, and do the job outdoors!

One other item of equipment that every serious bullet caster should have is an LBT bullet hardness tester. This little gadget specifies bullet hardness in BHN (Brinell Hardness Number) from 5 (pure lead) to 40 (copper) in just a matter of seconds. Although the hardness of paper patched bullets you shoot at a piece of paper may be of little significance, the hardness of the same bullet in game can mean the difference between a quick kill or a suffering cripple. There may be other bullet hardness testers on the market equally as good as the LBT, but some of them show comparative hardness on a scale of 1 to 10 instead of a scientifically recognized hardness number.

The Coleman Model 508 camp stove with a new Lyman melting pot containing pure lead. The bullet mould was made by Dave Farmer of Colorado Shooter's Supply. With this outfit, the author has cast thousands of paper patched bullets.

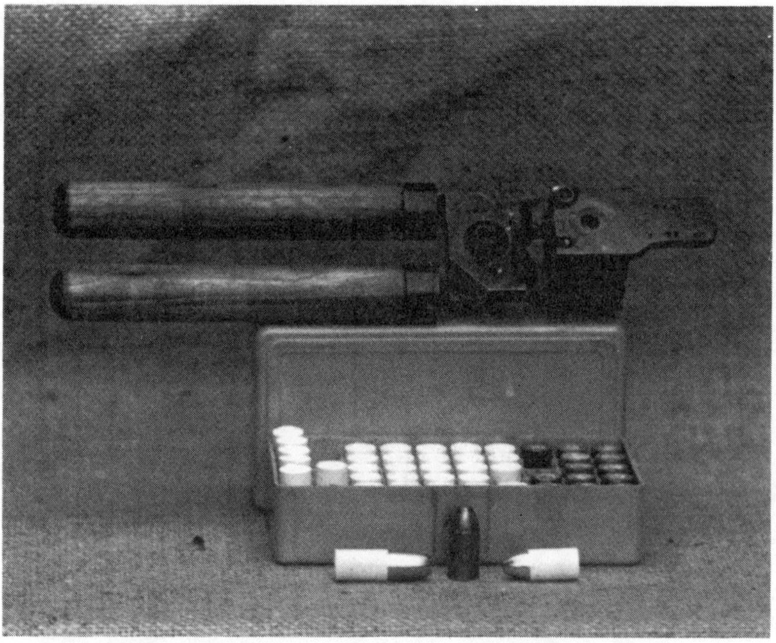

The Hoch mould made by Dave Farmer along with some 415-grain paper patched bullets for use in the .45-70.

4

Sizing, Bumping and Swaging

When we think of paper patched bullets, we give little consideration to anything except casting and patching. After all, that's the way it was done in the old days. Like the grooved lubricated bullet, however, there are a lot of things that can be done between casting and patching that will refine the bullet's performance.

Before we get into the details of bullet refinement, let's define what we mean by sizing, bumping and swaging, because these are different terms that mean different things to different people. I once carried on a half-hour conversation with a man about swaging bullets before I realized he was talking about sizing and lubricating in a Lyman No. 45 Sizer-Lubricator and I was talking about forming bullets from lead wire or cast slugs of lead.

The term "sizing" as used here means to *reduce* the diameter of a bullet, patched or unpatched, by running it through a die. The term "bumping" is used here to indicate an operation by which the bullet's diameter is slightly *increased* by inserting it into a die and applying pressure on one end of the bullet or the other to squeeze the bullet out to a larger diameter. "Swaging" is the operation by which we use extreme pressure in a press or pound die to force a cast slug, slug of lead wire or a cast bullet into the shape of a bullet, or to reshape it into another bullet by changing the ogive, adding a hollow point and/or reshaping the base.

Probably the most common operation, and perhaps the most versatile as far as the paper patched bullet is concerned, is that of sizing. It's common because practically everyone who gets interested in paper patched bullets is already enrolled in the school of grooved lubricated bullets and has a sizer-lubricator at their disposal.

Before I had a good .30-caliber paper patched bullet mould, I used very soft 190-grain bullets cast from my RCBS 30-180-FN mould. The bullets were first wiped with a thin case-sizing lubricant, then pushed through a .309-inch die in a SAECO sizer-lubricator. The lube was then wiped off and the bullets patched with nine-pound onionskin, bringing them up to a diameter of .316 inch. Loaded in front of 32 grains of IMR-4064, these bullets had an average velocity of 1,932 fps with an extreme spread of only 27 fps for five shots. Although I did not get the accuracy I wanted from this load, its ballistics were very close to the original Krag load, but with a more reliable bullet.

The load that I use most with this bullet is 19 grains of SR-4759 which delivers superb accuracy at 100 yards at a velocity of around 1,600 fps. This same load worked well with the conventional grooved lubricated bullet.

Not all .30-caliber chambers or throats will accommodate a patched bullet .316 inch in diameter, and even when they do, sometimes a hard ring of paper remains in the throat and interferes with the next round as it is chambered. To prevent this, the next logical thing to do is to lubricate the patched bullet and run it through the sizing die a second time. This can be a bit tricky with the skinny .30-caliber because, if the bullets are soft, they have little structural support and sometimes the force needed to start the bullet into the die will distort the nose. However, I have used this procedure many times and it is a good way to go if you want to take a stab at paper patching.

I should point out right now that the lubricant used in this procedure should be quite thin. I have used Rooster Laboratories' CSL-71 case sizing lube as well as my own mixture of 45 percent beeswax and 55 percent Vaseline™. Either will work, although I usually use my own, especially on larger diameter bullets.

Obviously there is a significant advantage to sizing paper patched bullets. It lets you use most of your existing bullet moulds as either grooved lubricated bullets or as paper patched, without the expense of going to a new mould. Far more significant is the fact that the paper patched bullet can be made much softer than its grooved counterpart and still be driven at velocities up to 2,200 fps or slightly faster with no loss of accuracy and an astronomical gain in expansion. This is a significant asset in the hunting field.

The .375 H&H is an excellent cartridge for handling grooved bullets that are sized down for paper patching. Bullets are large enough in diameter to have sufficient rigidity to withstand the sizing operation, and are usually accurate at right around 2,000 fps. Any time you hit a deer with a soft 300-grain bullet at 2,000 fps, you have venison in the freezer if you put the bullet anywhere near where it should be.

In my Ruger No. 1 I've used the Lyman No. 375110 shallow groove bullet (a bullet I designed in 1958) cast of lead with 3 percent tin for a weight of 302 grains. When patched with onionskin paper, lubricated and pushed through a .376 inch sizing die before loading over 47 grains of IMR-3031 with a Cream of Wheat filler, this bullet gives good accuracy at 1,980 fps which translates into 2,627 foot-pounds of muzzle energy. This is a superb load for any deer that ever walked!

Probably the best sizing die on the market is one produced by Dave Corbin of White City, Oregon. Dave catalogs them as reducing dies and they are approximately two-thirds the price of a custom adjustable paper patched bullet mould. One of Dave's reducing dies, however, works far better than the bullet sizing die in a sizer-lubricator and will serve another purpose that we haven't even mentioned.

First, the Corbin reducing die will fit in any ⅞-14 thread loading press and is used to reduce the size of grooved bullets from their as-cast diameter down to the bore diameter of the rifle involved. Using this die will double or triple the versatility of every grooved bullet mould for a given caliber. For example, during the 40 years I've used the .45-70, I've accumulated a number of bullet moulds for grooved lubricated bullets of that caliber. Many of these bullets are extremely accurate and some, like the Lyman 457132 Postell bullet, are special-purpose bullets designed nearly a century ago. In a very soft alloy, these bullets usually come from the mould measuring .458 to .459 inch, while from a hard alloy they usually measure .460. By using my Corbin reducing die, I can take any .45-70 bullet I have and reduce it to .452-inch diameter for paper patching. Again, just as when ordering a mould, I ordered the die with diameters based on pure lead. This means that my soft bullets come out of the die measuring .452 inch, and hard bullets containing antimony come out a trifle bigger.

Many times I have read that one cannot reduce the size of a cast bullet by more than .001 or .002 inch and maintain concentricity. Not so with the Corbin reducing die. This is a precision die that accepts the entire bullet at the bottom end, pushing it through a gentle taper to eject it out the top perfectly round and concentric and at the size you want.

When using the Corbin die, I lubricate my bullets with Rooster Laboratories' CSL-71 case sizing lube, and push them through the die nose-first with a perfectly flat punch furnished with the die. The punch replaces the shellholder in the loading press.

One of my favorite combinations using this procedure involved the 475-grain Postell bullet reduced to .452 inch and then patched with onionskin paper. When loaded over 60 grains of IMR-4064 and fired from my Ruger No. 1, this massive slug averaged 1,772 fps with about 3,300 foot-pounds of muzzle energy—a bear load if ever I saw one!

While discussing the Postell bullet, I want to mention that when patching grooved bullets, I always carry the leading edge of the patch just ahead of the first driving band. This helps prevent damage to the patch when the bullet initially engages the rifling. If the patch is applied wet as it should be, it will shrink tightly to the forward section of the bullet which is usually smaller in diameter than the driving band even after a trip through the reducing die. Plenty of lube on the bullet also helps protect the patch on its initial engagement with the rifling.

In patching grooved bullets there has always been the question of whether or not to use a gas check on those bullets having a gas check shank. Other than to say that I prefer a full-diameter base on my bullets, I cannot give any concrete proof that the addition of a gas check on paper patched bullets makes them any more accurate. Indeed, if the gas check is left off, then the bullet is usually a bit easier to seat in the case, unless you are using a tapered neck plug which we will discuss later.

My own preference is to use a crimp-on gas check, although I do it a bit differently. First, I anneal the gas check by holding it in a propane flame with a pair of tweezers until it turns red. Then with the gas check held against the bullet's base by the reducing die punch, I push the bullet through the reducing die. The gas check sizes down beautifully and doesn't have any spring-back to loosen it from the bullet shank. The paper patch is then applied with the assurance that the bullet has a perfectly symmetrical base.

A few pages back I mentioned an additional purpose of the Corbin reducing die that required discussion. This involves the use of paper patched bullets over black powder or Pyrodex. As mentioned in Chapter 2 due to powder fouling left in the throat by black powder or Pyrodex, the diameter of the bullet used in such loads must be smaller than that used over smokeless powder. One of the simplest ways to handle this situation without purchasing separate moulds for both black and smokeless, is to take a *patched* bullet of the proper size for smokeless powder and run it through a Corbin reducing die, thus reducing its size to the rifle's bore diameter. This works fine for me, and since I use both types of propellant quite extensively, the Corbin reducing die allows me to cast, patch and stock only one diameter of bullet for my .45-70s. When I need the smaller diameter for black powder or Pyrodex, I merely run the bullets I need through the Corbin reducing die.

When compared to the sizing operation, bumping a bullet to a larger diameter does not offer quite as many options. Where a bumping operation is needed, however, it is darned difficult to find a suitable substitute. Just using a thicker patch won't always do the job despite what you may have read and what would seem a logical conclusion. When using a smokeless load, the diameter of the unpatched bullet relative to the bore diameter is vital to accuracy and performance.

Bumping a bullet was first discovered in the ranks of the Cast Bullet Association, where certain members found that if the bullet nose's diameter was as large as the rifle's bore diameter, it helped align the bullet in the barrel, thus resulting in better accuracy. Therefore, in order to obtain the larger diameter nose without going to a custom mould, the bullet in question was inserted full length into the sizing die and a quick thrust of pressure exerted against the nose punch. This expanded the nose's diameter by .001 or .002 inch while leaving the bullet's body or shank the diameter of the sizing die. It was effective even on fairly hard bullets, with significant results on the target.

With paper patched bullets, the objective is to increase the bullet's body diameter without changing the shape of the nose or ogive, and without creating a shoulder at the juncture of the body and ogive.

Any good machinist can make a bumping die from mild steel similar to that shown in the illustration at the end of this section. The die body should be at least twice as long as the longest bullet to be bumped and the base-forming punch should be as long as the bore so that it can be used to eject the bullet. To ensure that there is no step-down shoulder at the juncture of the bullet body and ogive, the nose-forming part of the die set should be much larger in diameter than the die body's bore. By doing it this way, the machinist can make the ogive's origin match the bullet's body diameter. The final product will show a nice, smooth-sided contour. In order to prevent warpage, especially of the base forming punch, it is best to leave the parts unhardened.

In the actual bumping operation, the bullet should be lightly lubricated with a thin case sizing lube such as Rooster Laboratories' CSL-71. Make certain that the entire bullet, except the base and meplat, is covered with a thin film of lube. Without the lube, the bullet is apt to stick in the nose-forming cavity.

Once you have the die set assembled with a bullet to be bumped, set the die on a flat billet of steel or other hard, firm surface, and strike the end of the base-forming punch with a four-ounce, plastic-tipped hammer. Usually, two sharp raps are adequate to bump a bullet of pure lead and three raps will take care of a bullet as hard as 9 BHN. By using a plastic-tipped hammer, the base-forming punch will not be battered over and will not require hardening.

After bumping each bullet and before proceeding with the next one, be sure to remove any lubricant that might be on the inner face of the nose former or the matching face of the counterbore in the die body. During the bumping operation, the lubricant has a tendency to collect on these faces and gradually will build up sufficiently to hold the members apart and leave a ridge around the bullet.

Although the bumping operation may seem like a slow, tedious process, the results are worth the minor inconvenience.

I cannot overemphasize how vital the relationship is between the unpatched bullet's diameter and the rifle's bore. As mentioned in Chapter 2, when I tried to use a bullet having an unpatched diameter .0015 inch smaller than the bore diameter of my rifle, hardened bullets did not work at all with smokeless loads. Even pure lead bullets performed only when the combination of bullet weight and pressure was exactly right.

Now we come to swaging. Believe me there is no comparison between a swaged bullet and a cast bullet. The swaged bullet is smoother, more dense and more uniform in shape. It is a superior bullet in every respect, although not so superior as to eliminate all bullet moulds and the way we've been doing things for well over a century. If I were starting out in the paper patched bullet business today, I would invest in a good bullet mould to which I could later add bumping dies and/or reducing dies and finally, perhaps, a good swaging press and swaging die.

Bullet swaging adds an entirely new dimension to homegrown bullets. Although the bullet shape remains constant with the die just as it does with a mould, the nose and the base are changed easily by the use of different nose- and base-forming plungers. This is a significant asset.

My first real look at a swaged paper patched bullet was when Fred Cornell showed up at the house with a double handful of bullets he had swaged on his RCBS press with a die he had made. These bullets weighed about 550 grains and had been made for his muzzleloading .45-caliber slug rifle. Some of the bullets had a large diameter meplat, while others had a deep hollow point. Some had a perfectly flat base, while others had a bevel around the rim of the base and/or various-shaped base cavities ranging all the way from the usual cupped base to a deep truncated cone cavity that moved the center of gravity well forward in the bullet. The entire display was a real eye-opener to the number of things that could be done by swaging.

Swaging requires a lot of pressure, far more than I care to subject my 40-year-old Hollywood press to. In fact, even with Fred's heavy RCBS press, he was able to put a slug of pure lead in the die, push the handle down to its extreme position and then stand back and watch a thin wire of lead grow in length as it was extruded from the die's bleed hole. The fact that this extruded wire continued to lengthen for several seconds after the handle was in the extreme down position, meant that the frame of the press had been put under extreme tension and was gradually returning. This is not a criticism of a fine press, rather it illustrates the tremendous forces involved and suggests that if you go to bullet swaging, it would be wise to consider a press made specifically for that purpose. Dave Corbin of White City, Oregon, is the specialist in this field and has for years offered bullet swaging presses and dies for whatever the customer can dream up.

Fred Cornell made a press for me along with a swaging die, two base-forming punches and two nose punches. The die forms a beautiful paper patched bullet possessing an unpatched diameter of .452 inch and a long sweeping ogive terminating in a $\frac{9}{32}$-inch diameter meplat. A bullet weighing about 415 grains has $\frac{1}{2}$-inch parallel sides and is patched for $\frac{19}{32}$ inch of its length. This permits use of a bullet this heavy with 53 grains of IMR-3031 in a chamber that has no freebore, and it allows full utilization of case capacity when using black powder or Pyrodex. It is a beautiful bullet.

One of the nose-forming punches Fred made for this die allows me to form a $\frac{3}{8}$-inch deep hollowpoint that is .150 inch in diameter with a slight taper to the bottom. Hollowpointing the bullet serves three purposes. First, it lengthens the bearing surface of the bullet without increasing weight. Secondly, with a pure-lead hollowpoint bullet backed by 54 grains of RS Pyrodex for an average velocity of 1,300 fps, expansion on light-framed deer is virtually guaranteed. Thirdly, and perhaps most important, swaging a hollowpoint in the bullet ensures elimination of any possible void that could be present as a result of using a bullet cast from a nose-pour mould.

There are a number of different ways one can go in swaging bullets. You can swage bullets from extruded lead wire thus guaranteeing ahead of time that there are no voids in the bullet; you can swage from cylindrical slugs cast in a special mould; or you can take bullets from a nose-pour Hoch bullet mould and reform them in the swaging press. Although Fred Cornell enjoys working with a cast slug on the theory that the more working of the metal, the less chance of a void, I elect to cast bullets in my Hoch mould and then reshape them by swaging. Whichever way you choose, you won't go wrong.

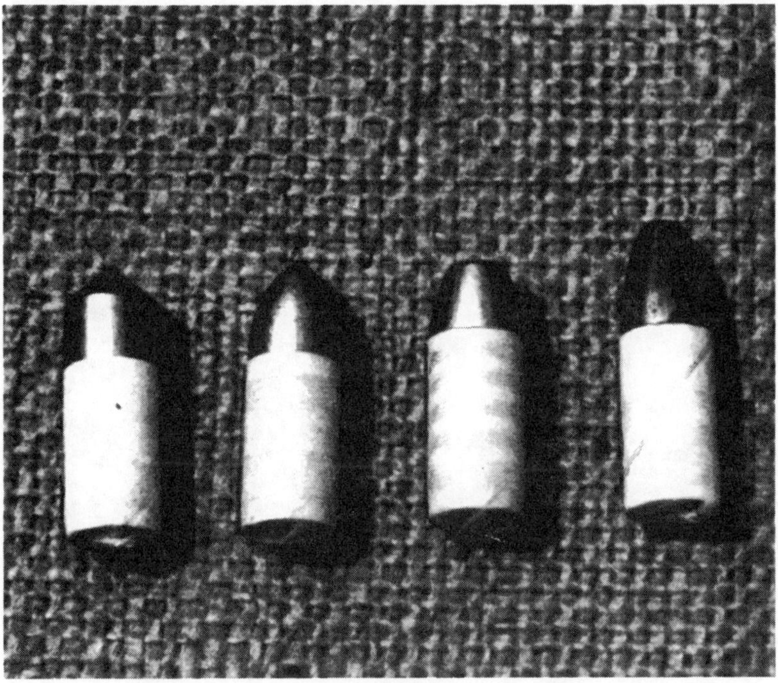

Close-up of Lyman grooved bullets after patching. From left to right, No. 375110, No. 375118 and No. 375449. At the extreme right is the traditional smooth-sided, cupped-base, paper patched bullet for the .375 from a mould made by Charlie Canoll.

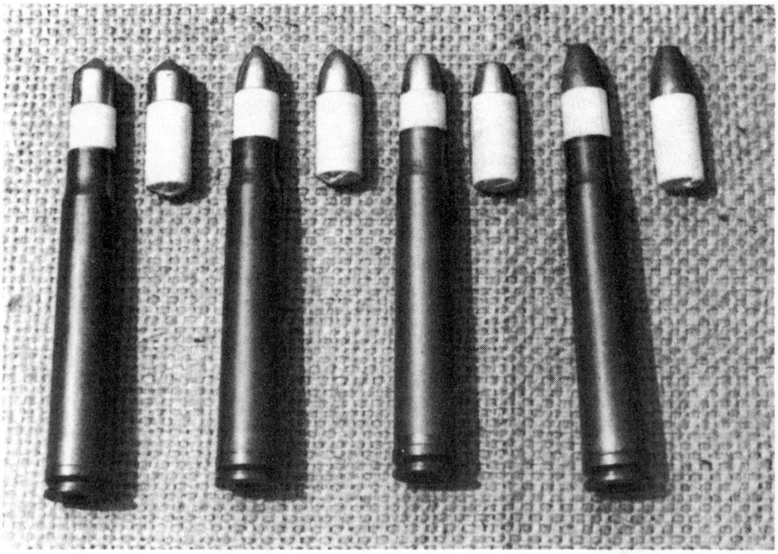

Four .375 H&H cartridges loaded with paper patched bullets for a Ruger No. 1. Left to right are the Lyman grooved bullets No. 375110, No. 375118 and No. 375449. On the far right is the traditional paper patched bullet.

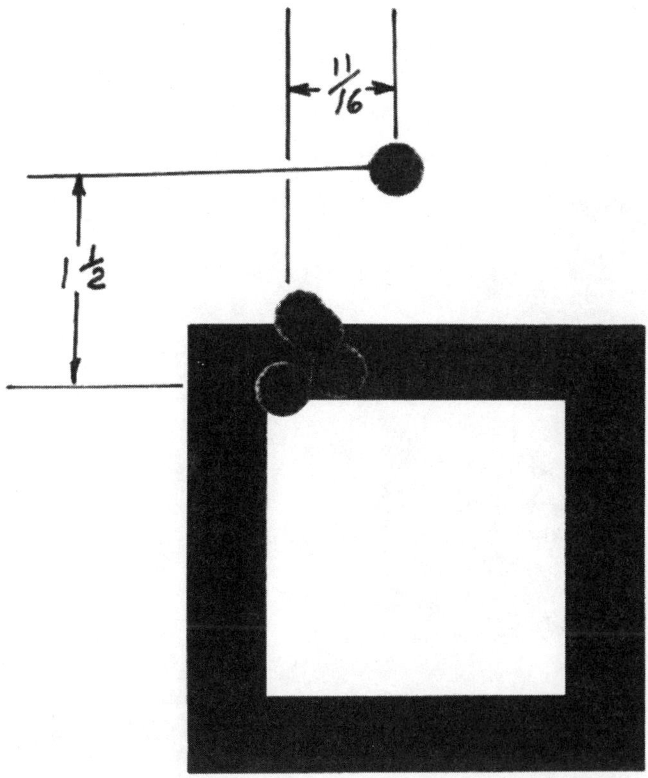

This group was made by the Lyman No. 375110 bullet weighing 302 grains and patched with nine-pound onionskin before being pushed through a .376 inch sizing die. The load was 47 grains of IMR-3031 with a Cream of Wheat filler for a velocity of close to 2,000 fps.

This shows the Corbin reducing die (inset) mounted in an old Hollywood press for reducing the as-cast diameter of .45-70 bullets down to .452 inch so that they can be paper patched. The bullet shown is the Lee No. 457-405F.

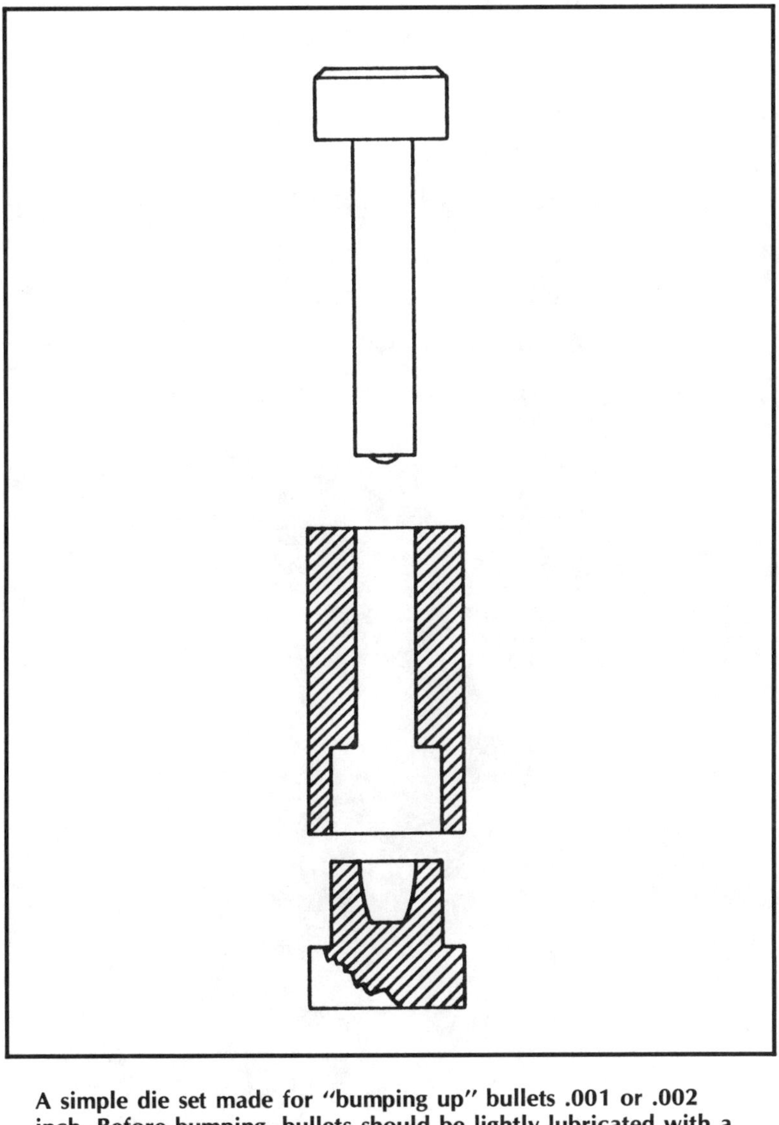

A simple die set made for "bumping up" bullets .001 or .002 inch. Before bumping, bullets should be lightly lubricated with a thin case sizing lube or similar lubricant.

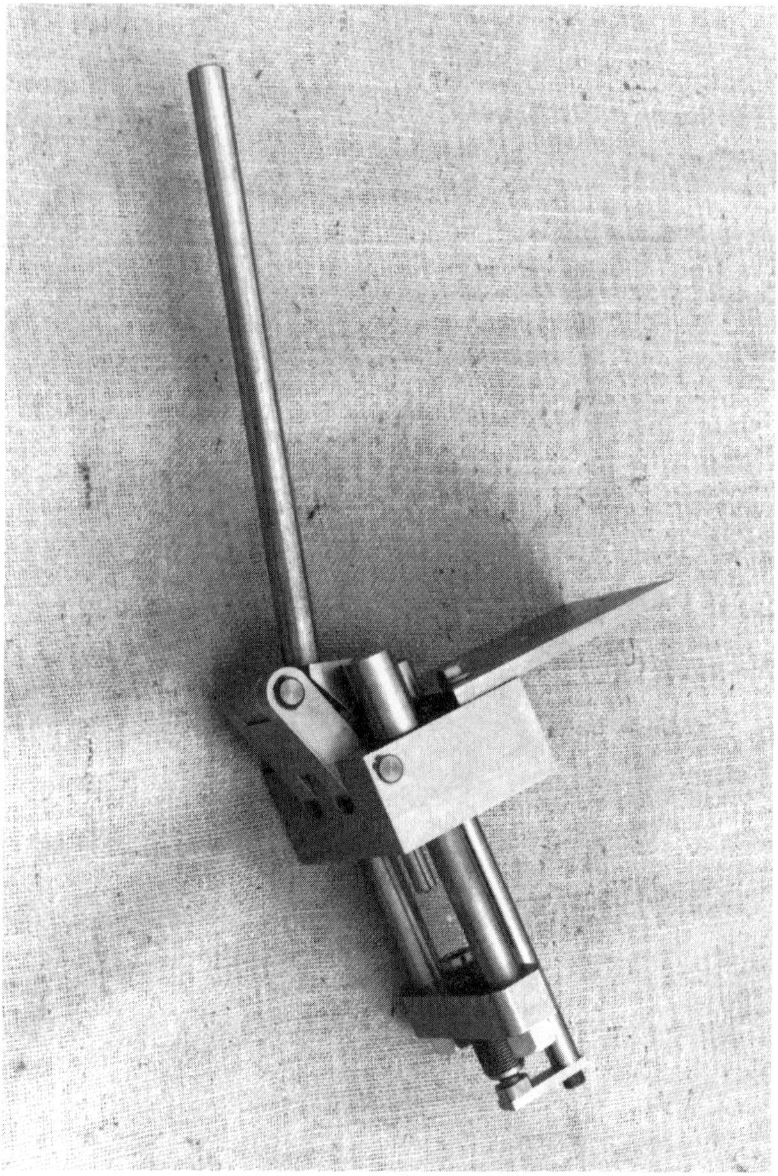

A rugged, high-leverage swaging press and die made by Fred Cornell for swaging paper patched bullets.

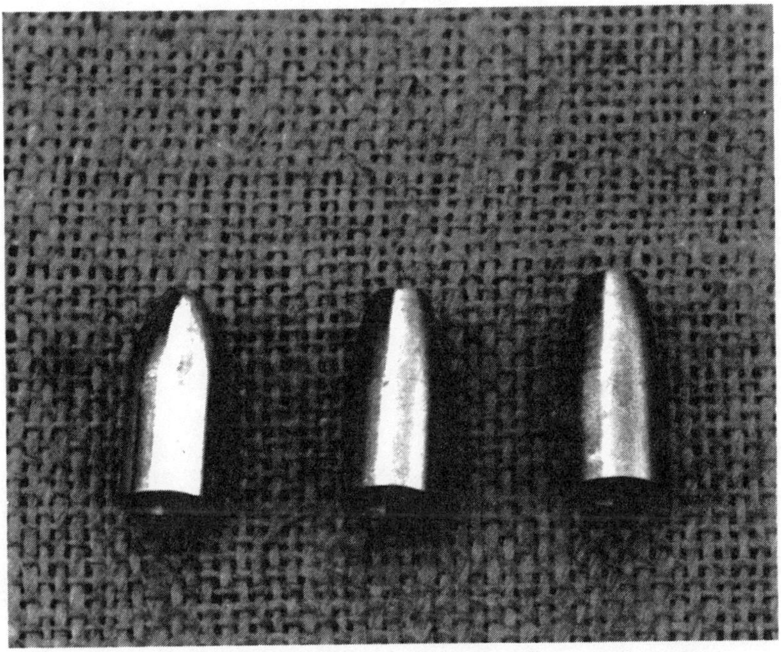

Three paper patched bullets of identical diameter and weight. The bullet on the left was cast in a Hoch mould, then it was swaged and reformed to produce the bullet in the center. It was then swaged a second time to form the hollowpoint bullet at the right. Note that both of the swaged bullets are longer than the original Hoch bullet, and that the hollowpointed bullet at the right is longer than either of the other two.

5

The Paper Jacket

Although the title of this book and the heading of this chapter might sound like the title of a mystery novel, the only mystery here (as I see it) is why handloaders are so slow to grasp the significance of the paper jacket. Because that is exactly what a paper patch is—a paper jacket, nothing more, nothing less. Had the paper patched system continued in popularity after the introduction of the metal jacketed bullet, it most likely would have been referred to as a paper jacket. The term *patch* was a carryover from the roundball muzzleloading days, and even the first metal jacketed bullets were marketed as *metal patched*.

Although old Bill Shakespeare would say that "that which we call a rose by any other name would smell as sweet," he never tried to sell roses under the name of skunk cabbage! As any writer knows—and Shakespeare knew it too—the written word evokes an image in the mind of the reader. When we read about paper patched bullets, we immediately conjure up images of a prairie blackened with buffalo and some grizzled old-timer out there with a Sharps rifle or Remington rolling block giving them what for amid clouds of black powder smoke and the solid "thunk" of a heavy lead slug striking flesh and bone.

There is no argument against this picture. I love the image myself and wish I could return there to participate in it, but that is only half of the picture. The other half is taking the same bullet and boosting its velocity anywhere from 300 to 700 fps with a liberal dose of smokeless powder. When we do that, we start thinking in terms of "jacketed" bullets instead of "patched" bullets. So while I don't want to dim or destroy a term that is a significant part of our heritage, I do want to emphasize that the term *paper patched* in modern thinking does not do justice to one of the greatest developments in firearms history.

If you doubt the significance of the paper (jacket) patch, consider this for a moment. The paper patched bullet leaves no metal fouling in the barrel—no lead and no wash of jacket metal. In fact, you can shoot paper patched bullets all day long with smokeless powder and easily clean the rifle at night with two or three passes with a dry patch. Secondly, the paper patch in no way restricts expansion of the bullet regardless of velocity. Expansion is governed solely by the alloy and the bullet nose shape. Try poking a deer with a metal jacketed bullet at 1,300 fps and see how much expansion you get! Some maybe, but not much compared to the bullet that is unencumbered with a metal corset. Once again, in this age when the home-grown alloy bullet is reappearing in the game fields, try firing a grooved lubricated bullet cast of lead and 3 percent tin at 1,800 to 2,200 fps and then tell me about the absence of accuracy and the proliferation of lead in the barrel. The paper jacket takes care of all this!

In all honesty I should point out that the new Marlin 1895SS .45-70 with the improved deep-groove Micro-Groove barrel will handle soft lubricated bullets at 1,800 fps without any leading. This, however, is an exception and not the rule.

Originally, there were three distinct types of paper patches, two of which were used only on the target range and the third which served the targeteer, the military and the average hunter-shooter.

One of the patches used only for target shooting was the Chase patch. This patch was a simple rectangular or square patch of the desired thickness which was just long enough to encircle the bullet once with its square ends abutting one another. The patch was used dry and applied to the bullet at the target range during the loading process. Bullets using the Chase patch could not be loaded into a cartridge case, but were loaded directly into the barrel by use of an Ideal Ball Seater sold by the old Ideal Manufacturing Company.

The patch was first rolled into a tube and then inserted into the Ball Seater with the tube's end projecting a bit beyond the dummy shell's mouth. The patch was squared by gently pressing it against the plunger and then the bullet was inserted within the patch and seated against the plunger. The Ball Seater was then inserted into the rifle's chamber and the plunger was pressed forward to seat the bullet and patch in the barrel.

As awkward as this may sound, the Chase patch gave exceptional accuracy. On July 18, 1891 Mr. Francis J. Rabbeth, using a .38-55 with 55 grains of Fg Hazard powder and Chase patched bullets cast of one part tin to 25 parts lead, fired 15 consecutive shots into a two-inch ring at 200 yards. With anything less than a good bench rifle, one would be hard put to duplicate that performance today with a modern rifle, scope and jacketed bullets.

The second type of paper patch was used only with rifles having a false muzzle. Here the patch consisted of three strips of paper placed across the false muzzle and each strip crossing the other at its midsection at a 60 degree angle. Sometimes two strips were used at right angles. The bullet was then placed on the patch's midsection, directly over the bore and pressed home to a determined depth. Again, as with the Chase patch, the strip patch was used only on the target range.

The patch we are interested in is a wraparound patch; a patch shaped like a parallelogram with its ends cut at a 35- or 40-degree angle and just long enough to wrap around the bullet twice without overlapping. On rare occasions I have used a patch that would go around a bullet three times, but twice is the norm. Unlike the Chase and strip patches, which are applied dry, the wraparound patch is applied wet or damp so that as it dries it shrinks tightly against the bullet, thus making a paper jacket. A properly applied wraparound patch is so tight to the bullet that it cannot be twisted or turned on the bullet after it is dry. While the Chase patch and the strip patch had to be applied during the loading process, the wraparound patch was applied prior to the loading process and was the only paper patch system that could be used with fixed ammunition.

There was one fault with the wraparound patch (though in my mind I believe it was more a fault of the rifle than the patch) and it gave rise to at least one myth that currently exists. The fault of the wraparound patch was that occasionally the patch failed to leave the bullet at the rifle's muzzle, thus causing an off shot. This was intolerable for the target shooter and led to the development of the Chase and strip patches which left the bullet at the muzzle without fail. Shooters who used the wraparound patch made every effort to assure that the patch left the bullet by wrapping the patch in the opposite direction of rifling twist. Their theory was that this would unwrap the patch at the muzzle. It worked most of the time.

It is my opinion—and strictly an opinion—that this problem existed due to the shallow groove rifling in most rifles of that day which were built for paper patched ammunition. I support this theory for two reasons. First, the paper patched bullet is gripped easily by the rifling and rotated, even though the rifling barely marks the bullet. I have fired and retrieved paper patched bullets cast of type metal that were only .0005 inch larger than the bore diameter of the rifle, and the marks of the rifling were barely discernible. Secondly, knowing this to be a fact, and with cost of production always uppermost in mind, rifle manufacturers saw no need to waste time and money on deep-grooved rifling when shallow-groove rifling would do as well. Thus was borne the myth, at least in the U.S., that rifling for paper patched bullets was best if it was shallow.

The British, and perhaps most of the other countries of Europe, did not necessarily accept the shallow-groove theory. The 577/450 Martinis had very deep rifling and according to some writers would handle bullets ranging all the way from .450 to .470 inch in diameter. I can't say whether the patch was sliced or unwrapped in these barrels. The point is that shallow rifling is not a requirement of good performance for paper patched bullets and was not recognized as such in all areas. As another example, in the March 1984 issue of the *Black Powder Report* writer Gareth C. Ward wrote of obtaining a German Model 71/84 chambered for the .43 Mauser. Once again, the military ammunition was loaded with a paper patched bullet, although the groove diameter of the rifle in question was .446 inch and the bore diameter was between .439 and .440 inch—nearly the same depth as today's rifling.

Finally, on page 19 of *The Breech-Loading Single-Shot Rifle* by Ned Roberts and Ken Waters, Ned Roberts writes:

> Ballard rifles were rifled with deeper grooves than either the Remington or Sharps, and were chambered so that the throat permitted the breech-seating of either lubricated or paper-patched bullets of proper size.

Again, speaking of the Ballard on page 47:

> Paper patched bullets remained the favorite while black powder was in use, but with the coming of smokeless powder it was found that grooved, lubricated bullets gave superior accuracy. Of course, as previously mentioned, the deeper rifling of Ballard barrels was also of the greatest importance in securing accuracy with lubricated bullets.

The point of Ned Roberts' comments are that although shallow grooves were *not* necessary for good accuracy from paper patched bullets, deep grooves *were* necessary for good accuracy from lubricated bullets.

From my own experience, it is far more desirable for the rifling to slice the patch in longitudinal strips that leave the bullet at the muzzle than to risk having the patch cling to the bullet. This, of course, requires deeper rifling than was used in those rifles made only for paper patched bullets.

In fairness to the reader, I must point out that this opinion is diametrically opposed to that of Frank H. Mayer who, in the December 1934 issue of the *American Rifleman* wrote:

> When lubricated, grooved, naked lead bullets were used, these worked best in deeper-cut rifling, for which the paper patched projectiles were unsuitable. The shallow-grooved barrels best adapted to paper patched bullets and having grooves only .002 to .0025-inch deep in .40 and .45 caliber, would not properly spin the naked grooved and lubricated bullets, which stripped badly in the shallow rifling. Conversely, the paper patched bullets were entirely unfit for the deeply rifled barrel, the deep grooving—from .003 to .005-inch in depth—cutting and destroying the patches, with unsatisfactory ballistic results.

Although I recognize that Frank Mayer was knowledgeable of that which he wrote, I also submit that the British and European arms and ammunition manufacturers put more into the development of the paper patched bullet than was done in the United States. Further, having spent over 27 years as a technical writer for industry, I know how easy it is to turn an engineering cost reduction into a consumer asset through the manipulation of words.

In the heyday of paper patched bullets, patches were cut from a special linen fiber paper referred to as bank note paper. This was available in four thicknesses: extra thin (.0015 inch), thin (.002 inch), medium (.0025 inch) and thick (.003 to .0035 inch). By using the various thicknesses of paper, one could regulate the bullet's final diameter to match the rifle bore. In cases where two or three different bullets were being used, the difference in bullet diameter as they came from the mould could be compensated for by using a different patch thickness.

Frank Mayer also wrote in the December 1934 issue of the *American Rifleman* that he used different thicknesses of patching to compensate for the varied thickness in brass cartridge cases of different manufacture. According to his article he was not alone in this routine.

Basically, the paper patch does three things. First, it provides a lubricated non-fouling barrier between the lead alloy bullet and the barrel, and thus prevents leading. Secondly, it prevents gas cutting, the major bugaboo of the grooved, lubricated bullet. Thirdly, the paper patch helps seal the bore against gas leakage.

Those three functions of the paper patch all occur within the barrel. Once these functions are served the paper patch is of no further use and should be stripped from the bullet—that is, until such time as we can devise a way of keeping a fully intact patch on the bullet during its journey from the muzzle to the target.

Now over the years we have read about the original patches being made of linen fiber paper, and how those were a superior patch because of the paper's strength. Even today there are many shooters, myself included, who have paid a good price for 100 percent cotton fiber paper in order to have a stronger patch.

Like the shallow grooved rifling, I believe this is the wrong approach. When I fire a paper patched bullet, I want the patch completely free of the bullet the instant the bullet exits the muzzle. To have it any other way is detrimental to accuracy as evidenced by the development of the Chase and strip patches. Thus, in order to make certain my patches are sliced by the rifling and are free from the bullet as it exits the muzzle, I want to use the thinnest paper practical (usually nine-pound onionskin), and I don't want it any stronger than 25 percent cotton fiber.

A nine-pound paper is much easier to apply to a bullet than a 16- or 20-pound paper, and a 25 percent cotton fiber paper is much easier to apply than a 100 percent cotton fiber paper of the same weight. In both cases, the weaker paper makes a smoother patch and folds over the base of the bullet or twists in a tail better than a stronger paper. This advantage is because it is more limp when wet.

The next thing we often read about, and that has been passed down to us, is that the patch should be cut with instead of across the fiber, again in the interest of strength. Actually, it makes little difference which way you choose as long as all of your patches are cut in the same direction. Patches cut across the fiber stretch a bit more when wet than patches cut with the fiber, so if you cut across the fiber make sure the patches are $\frac{1}{32}$- to $\frac{1}{16}$-inch shorter.

I prefer to cut my patches across the fiber. They are more limp when cut from that direction, thus are easier to apply to the bullet and make a better fit when dry. More important, when we patch a bullet, we normally extend the patch a bit beyond the ogive's origin. This is done to ensure that all of the bullet shank that is larger than the rifle's bore diameter is covered with paper, and to help prevent the patch from stripping when it engages the rifling.

That portion of the patch that extends beyond the origin of the ogive, however, has to shrink more than the rest of the patch in order to snug the bullet. Because of this, patches cut with the fiber usually pucker in two or three places around the bullet where the patch overhangs the ogive, whereas patches cut across the fiber will shrink tightly and smoothly around the ogive.

In cases where the bullet is .0005 inch or so smaller than the rifle's bore diameter, it is not always necessary to extend the patch beyond the ogive's origin, but merely patch enough of the shank to provide at least a full caliber of bearing surface. I have done this when I wanted to use a long, heavy bullet and didn't want to compromise the capacity of the cartridge case.

Just how do we tell which way the fiber or grain runs on a sheet of paper? Quite often it is listed on the box, but many times it isn't. Although I've been told by those in the business that it is common throughout the industry for the grain to run perpendicular to the watermark, I have also been told that the direction of fiber varies with the paper being produced.

If the box or package is not marked or you are using an unknown paper, there is a simple test to determine the fiber's direction. Cut two strips of paper of the same length and width from the same sheet. Cut one strip across the short way of the sheet and the other strip the long way of the sheet. Hold these two strips by one end between your thumb and finger, and note which strip bends down the farthest. This is the one cut *across* the fiber. Strips cut with the fiber are much stiffer.

Controlling the patched diameter of the bullet with the patch's thickness often is necessary to make the bullet chamber freely in a tight throat or to make a bullet fit in an oversize bore. In short, by using different thicknesses of paper, you can tailor your bullets to a specific rifle.

Although it is no problem to obtain a thicker paper for patching purposes, thinner papers are sometimes difficult to find. Most of my patches are cut from a nine-pound onionskin that measures approximately .0025-inch thick and, at present, is readily available in the 25 percent cotton fiber. If a thinner patch is needed—and I have used them—go to a fabric shop and purchase a lady's dress pattern. The *Simplicity* patterns I have used are .002 inch thick, while *McCalls* and *Butterick* are .0015-inch thick. These papers are quite flimsy to work with and will not tolerate severe twisting and stretching at the base, but they do roll on nicely and shrink to a good, tight fit. Most important, they work in the rifle barrel.

Although not by intent, the computer industry has supplied us with an excellent source of paper for patching bullets. Some of the paper is of the groundwood or recycled variety, and some is high-grade letterhead 25 percent cotton fiber.

Paper used for data processing is green bar to make it easier to read long lines of figures, while plain white paper is used for word processing. Best of all, this paper is available in two-, three- or four-part for duplicating purposes, with the under sheets being of 12-pound weight and the over sheets about 20 pounds. This provides the opportunity to purchase a selection of weights and thicknesses in the same box.

When using computer paper, especially the multiple-part kind, it is best to moisten it with some medium other than saliva to avoid ingesting the chemicals used in the carbonized backing. Two egg whites stirred, not beaten, in a pint of water will do a good job. When you are finished, put the medium in a sealed glass jar in the refrigerator as it will keep for months.

Because of the way computer paper is used, running over sprockets on an automatic printer, it is tough and durable. In fact, some computer paper—often the under sheets in multiple-sheet packets—have a hard glazed surface that requires a bit more time to moisten to make certain it is done evenly and thoroughly.

It should be pointed out here that after a paper patch is stretched to go around a bullet and then shrinks to a tight fit, it is no longer as thick as it was before the patching process. This means you can't take a .003-inch thick patch, wrap it around a bullet twice and figure an increase of .012 inch. It doesn't work that way. Four thicknesses of 16-pound paper increase the bullet diameter about .010 inch, and four thicknesses of nine-pound paper increase the bullet diameter by about .008 inch.

Finally, let's discuss the physical size and shape of the patch itself. As for the size, there are no cast-in-concrete dimensions. Bullets vary in diameter; paper varies in stretch and shrinkage, and shooters vary in method of application. My experience, however, has been that for the .45-caliber the patch should be about 2 11/16 inches long; for .375-caliber, about 2 3/16 inches long; and for the

.30-caliber, about $1\frac{13}{16}$ inches long (this is with the patch cut across the grain or fiber). The finished patch should go around the bullet twice, leaving a slight gap between the ends of the patch on the final wrap. The ends should never overlap.

As for the width of the patch, cut it wide enough to cover that portion of the bullet shank that will contact the barrel, and leave enough overhang at the base so that when twisted and pressed into the cupped recess, it totally covers the bullet's base. I do not leave enough overhang to twist into a tail on cupped-base bullets, just enough to cover the base. Flat-based bullets are a horse of a different color. These require enough overhang to twist into a tail which is later clipped off.

In the June 1940 issue of the *American Rifleman*, Elmer Keith commented on possible problems with either the cupped-base or flat-based bullets. With cupped-base bullets and black powder loads employing the use of lube wads and card wads, quite often a wad is driven into the cupped base with such force that it adheres to the base, thus destroying the bullet's accuracy. Keith wrote that one of the ways to prevent this was by using a small amount of Cream of Wheat between the wad and the bullet's base.

I have recovered bullets from my sandbox in which the card and lube wads were driven into the cupped base, and I assumed that such bullets represented the flyers on my target. This condition is not so much of a problem with smokeless loads where it is unnecessary to use the wad materials.

In regard to the flat-base paper patched bullet, Keith said that unless the tail of the patch was cut off, it was driven into the bullet's base with such force that it made a sizable dent, again destroying the potential for fine accuracy. Once again, this should not be a problem with smokeless loads where there is no wad material present to sledge against the patch's tail.

It should be pointed out that even with smokeless loads where no wad material is used between the bullet and the powder, recovered bullets often show imprints of powder kernels that were slammed into the bullet's base at the time of ignition. Although this has to have an effect on accuracy, I am not certain that the use of a cardboard wad to protect the base does any better.

The patch's shape can be either rectangular or the more conventional parallelogram. I've used both, and one will work just about as good as the other. Often the rectangular patches produce more patches from a sheet of paper, but the parallelogram patches are tighter to the bullet along the patch's leading edge. This is because the angle of stretch coincides with the angle at which the end of the patch is cut. I have cut patches at various angles ranging from 25 to 45 degrees. As far as I'm concerned, 35 or 40 degrees is about optimum. I use 40.

Now I can't say that's all there is about a paper patch, because I'm certain it isn't. As time goes by, we all will learn something new—new materials and/or new patching methods. Some shooters already have used Teflon™ tape with good results, and I know of others who have experimented with such things as masking tape. Until I find something better, however, I'm going to stick with regular typing paper wet with saliva and wrapped around a bullet as best I know how. I'm also going to forget the necessity of a linen fiber patch and the desire to have a patch unwrap itself at the rifle's muzzle. That might have been desirable 100 years ago, but why? Did the linen fiber really make a better patch, or was it because the shelflife of the linen fiber exceeded that of the cotton fiber paper? Or possibly the linen fiber patch was less vulnerable to handling by the common soldier of every army in the world except the U.S. Whatever the reason, it is lost to us.

The author shooting his Navy Arms rolling block over cross-sticks. Although the days of prairie blackened with buffalo are long gone, such a rifle as this brings back the flavor of the old days when black powder and paper patched bullets were king. One of the best loads for this particular rifle is a swaged 480-grain paper patched bullet driven by 27 grains of SR-4759 and a Dacron filler.

The early metal jacketed bullets were marketed as "metal patched" as indicated on the label of this box of Winchester 255-grain .38-55 bullets. The label reads in part:

Winchester

Metal Patched

Grooved Bullets

Patented December 14, 1886

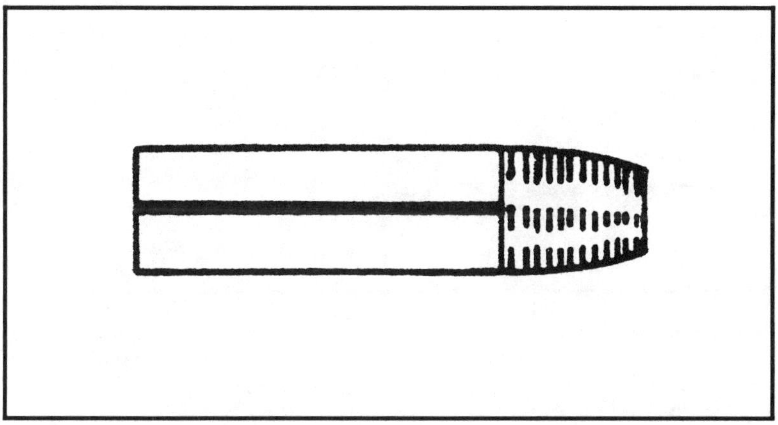

The Chase patched bullet with its single-wrap patch. When using a bullet patched in this fashion, it was also necessary to use the Ideal Ball Seater to seat the bullet in the barrel ahead of the loaded cartridge case.

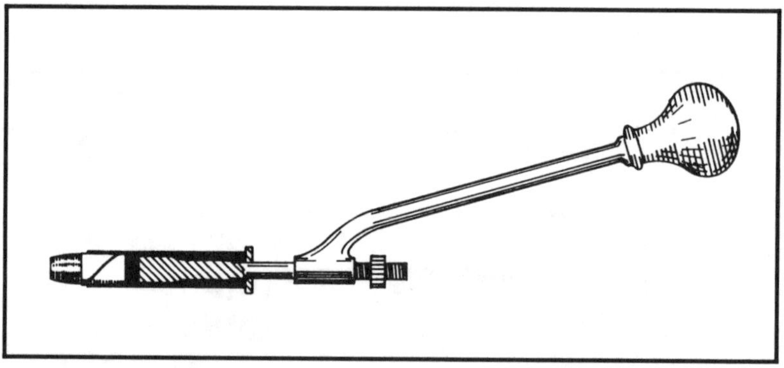

The Ideal Ball Seater could be used with the standard paper patched bullet, Chase patched bullet or a grooved lubricated bullet. Its function was to seat the bullet in the barrel ahead of a charged cartridge case. As shown in the illustration, the Ball Seater was essentially a dummy cartridge case with a plunger and handle. The bullet was placed in the dummy case and the entire unit inserted into the rifle chamber. A gentle, firm push on the handle seated the bullet in the barrel to a predetermined depth.

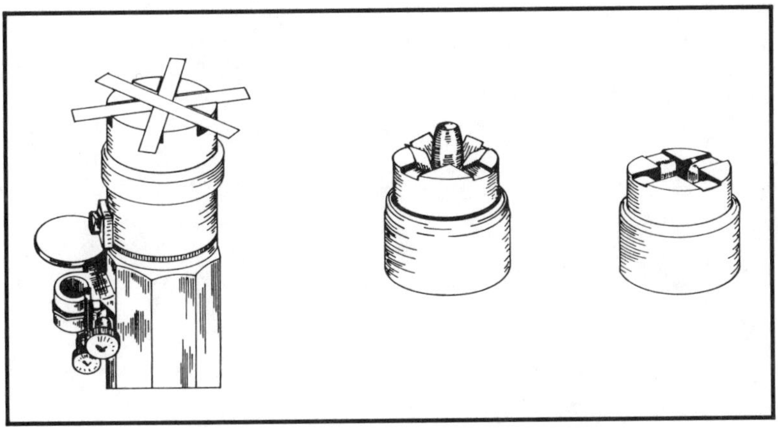

The strip patch consisted of either two or three strips of paper placed across the false muzzle of the rifle. The bullet was then placed directly over the bore and seated home with a loading rod.

6

Applying the Patch

No inanimate object draws the attention of a rifleman like the sight of a straight-walled case loaded with a thumb-sized paper patched bullet. For some reason, perhaps that white collar of paper around the bullet, such a cartridge evokes an image of power—visions of prairie blackened with thousands of buffalo. Whatever the reason, I suspect that at least part of it is due to the fact that paper patched bullets were always hand-wrapped with such precision they exuded an air of perfection. They were as perfect as human hands could make them, with the patch located at the same spot on every bullet and wrapped with an evenness that bordered on art. This is what we are after—bullets as perfect as we can make them and each one of them patched exactly like its predecessor.

In patching bullets, the first step is to determine the size of the patch—its width and length. In all probability you will eventually settle on one or two different widths for a given caliber, but for optimum performance, patch width really should vary with the length of bearing surface on a given bullet.

The patch for a cupped-base bullet should be wide enough to cover the bullet from a point about 1/16 inch beyond the ogive's origin to about two-thirds the bullet's diameter beyond the base. In short, for a cupped-base bullet you want enough overhang to completely cover the cupped recess, but not too much overhang to twist into a hard knotted tail.

Although this might sound contrary to what we have been led to believe, it should be mentioned that 100 years ago not all commercial bullets had the patch twisted into a tail long enough to be tucked into the base cavity. On page 34 of the November 1983 issue of the *Black Powder Report* there is a photograph of an original box of 330-grain .40-caliber paper patched bullets marketed by Marlin Fire Arms Company. This photograph clearly shows the cupped base and the fact that the patch overhang is not long enough to be twisted into a tail, but is merely folded over the base's rim. The patch does not even cover the entire base!

If by chance you have a bullet with a very deep cavity in the base, leave just enough overhang to turn over the base rim; do not leave enough to tuck way up into the cavity. Fred Cornell gave me some swaged deep-cavity bullets for testing purposes, and I found that when the patch was tucked full depth into the cavity, bits of the patch clung to the bullet destroying all semblance of accuracy. The bullets did look good though, and that deep cavity put the center of gravity well forward like a bar of soap in a wet sock!

The patch for a flat-based bullet should be wide enough to cover the bullet from a point about $\frac{1}{16}$ inch beyond the ogive's origin to about one-and-a-half times the diameter of the bullet beyond the base. This will be twisted into a tail and later clipped off nearly flush with the bullet's base. Since the patch will shrink to a tight fit, extending it a bit beyond the ogive's origin helps prevent tearing the patch when the cartridge is chambered with the bullet engaging the rifling.

It is easy to overdo extending the patch beyond the ogive, especially on a bullet having a long, sweeping ogive. When the patch is extended too far forward, it doesn't get sliced by the rifling and accuracy suffers.

One other thing that should be mentioned relative to patch width, is that if you are using a long heavy bullet that is just the bore diameter of the rifle or a hair under—say a .450-inch diameter bullet in a .4505 or possibly .451 bore—you can regulate the width of the patch to provide the desired amount of rifling engagement surface and seat the bullet way out in the case. This is the procedure I follow when using a .450-inch diameter 565-grain bullet in my .458 Ruger. My bullet is 1⅜ inches long and ¾ inch of that is covered with patch.

Personally, I like to patch as much of the bullet shank as possible, but never less than a full caliber and preferably closer to a caliber and a half.

In Chapter 5 I listed the lengths of the patches I used on .45-, .375- and .30-caliber bullets. These lengths were for patches cut across the fiber or grain, a direction that gives maximum stretch as well as maximum shrinkage. Since you might be working with a different caliber, or because patching techniques may differ from mine, we'll take a look at determining patch length.

There are two ways to determine the basic length of the patch: (1) by arithmetic and (2) by actual practice. Either way, however, is only a starting point.

Using the first method multiply the diameter of your bullet by 3.141 to get the circumference, then multiply this product by 2 to provide for a double wrap. For example, let's say the bullet is .450 inch in diameter. Multiplied by 3.141 equals 1.413, and multiplying that by 2 gives 2.826 inches, or about $2\frac{53}{64}$. If one subtracts $\frac{1}{32}$ or .031 inch from this so that the ends don't overlap, the result is 2.795 or $2\frac{51}{64}$ inches for a starting point.

The second way to determine the length of a patch is to wrap a strip of patching paper around the bullet three times. Wrap it firmly and square and then with a razor blade or utility knife cut through all three layers of paper near the base of the bullet. Unwrap the strip and note that there are three slits in it. The distance between the two outermost slits minus $\frac{1}{32}$ inch is your basic length.

Using this basic length as a starting point, cut two or three patches to the desired length and shape. *Important*: Make certain the ends are cut to a 35- to 40-degree angle so that when applied to the bullet, the seam angle of the patch will be far different than the pitch angle of the rifling. If you make the seam angle too shallow, it is possible that it will match the rifling's pitch and you will have one barrel land riding the seam on only one layer of patch. This will result in an unbalanced bullet and may strip the patch in the barrel. Obviously a rectangular patch will not match the rifling pitch.

With two or three patches cut to the predetermined size and shape, apply them to the bullets as described later in this chapter. Almost immediately you will notice the patches are far too long and the ends overlap. This is because after the patch is moistened, it stretches very easily, some of it during the moistening process and some during the wrapping process. You will have to cut-and-try to obtain the proper length.

Determining the thickness of paper to use is a little bit of "by guess and by gosh," unless you happen to be loading for an original rifle designed, made and chambered for black powder and perhaps for paper patched bullets. In that case, you will probably do as Frank Mayer did and patch the bullets to a snug fit in the fired, unsized cartridge case.

There are a lot of old .38-55 rifles around originally made for black powder and having inside barrel dimensions larger than the chamber. In other words, if you load a cartridge using a bullet large enough to fill the grooves, the cartridge will not chamber. This is the place to use black powder or RS Pyrodex and a putty-soft paper patched bullet that just seats in an unsized, fired cartridge case.

For use in modern rifles with modern chambers, I usually patch the bullet to fill the chamber's throat. When I first started with paper patched bullets, I used .003 inch thick, 16-pound paper which gave a patched diameter on the plus side of .460 inch with my .45-caliber bullets. Often I ran these through a .458-inch sizing die after the patching operation, but found that this reduces the diameter of the bullet under the patch. Today I patch most of my bullets for all calibers with .0025-inch thick nine-pound onionskin paper. With the bullet diameters I use, this usually results in a patched diameter about .001 inch larger than nominal groove diameter of the caliber involved.

Once in a while for reasons of which I'm not entirely certain, you will run into a rifle that is fussy about patch thickness and/or bullet diameter. A case in point is my Marlin 1895SS .45-70. This rifle has the improved Micro-Groove rifling with deep grooves, and it handles very soft grooved lubricated bullets with full-throttle loads almost to perfection. When I was using paper patched bullets in this rifle, I had to use a .450-inch diameter bullet patched with .003-inch thick, 16-pound paper. Whenever I tried a .452-inch diameter bullet patched with .0025-inch thick nine-pound onionskin, I picked up some leading in the barrel and accuracy suffered. I believe this was due to the fact that the Marlin barrel has 12 lands which stressed the paper a lot more than the normal six lands in my Ruger barrels.

Very briefly let's go through the steps of patching a bullet, and then return to elaborate on those steps where necessary.

1. Moisten the patch—don't soak it.
2. Lay the patch on the table with one end pointing directly toward you and hanging over the edge of the table about ¼ inch.
3. Lay the bullet on the patch so that the patch overlaps the ogive by approximately $\frac{1}{32}$ to $\frac{1}{16}$ inch. If you are using a step-sided bullet, place the bullet on the patch so that the patch is against the ogive shoulder.
4. Bring the point of the patch up over the bullet and press it lightly against the bullet.

5. While pressing lightly against the bullet, roll the bullet away from you, keeping the leading edge of the patch even.

6. If you are patching a cupped-base bullet, gather the overhang between your thumb and two fingers and, while twisting it in the direction in which the patch was applied, tuck it into the base cavity. Then, using the rounded end of a ball-point pen or dowel, press the overhang tightly against the cupped base, twisting the dowel in the direction in which the patch was wrapped. Finally, press the bullet's base against the table and give it a quarter turn in the direction in which the patch was wrapped.

7. If you are patching a flat-based bullet, twist the overhang into a tail, twisting in the same direction in which the patch was wrapped. Be careful not to tear the patch around the base's sharp edge. (Some of the old-timers did not leave enough overhang for a tail, but merely enough to be folded over the base's rim, leaving the center of the base uncovered. If you do this, then press the base of the bullet against the table and give it a quarter turn in the direction in which the patch was wrapped.

8. Stand the bullet nose-down in a loading block to dry.

9. With flat-based bullets having a patch twisted into a tail, clip the tail off close to the base *after the patch has dried*.

Those are the basic steps in patching a bullet. Now let's elaborate on those steps requiring more detail.

1. There are different ways to moisten patches and different mediums in which to moisten them. My preference is to moisten the patch with saliva by putting one end in my mouth and drawing it out between my lips and then reversing the patch and repeating the process. This gets the patch just about the right dampness, and saliva has about the right consistency to make the patch stick to the bullet. There are others who draw the patch

across a bowl of water, and still others who moisten a number of patches between damp towels or who draw the patch across a wet sponge. Whatever method suits your preference is the way to go. The important thing is to totally dampen the paper so that it is limp, but do not soak it, as this will cause it to tear during the patching process.

One other medium you can use to dampen the patch is the egg white/water medium mentioned on page 64. Whatever you do, don't use mucilage or glue on the patch; the patch must leave the bullet cleanly at the muzzle.

2. The direction in which the patch is wrapped can have a significant effect on the bullet's accuracy. Some of the old original rifles made especially for paper patched bullets had extremely shallow rifling—Frank Mayer reported about .002 inch. In these rifles, the patch is not sliced by the rifling, but is unwrapped from the bullet as it leaves the muzzle. For use in these shallow groove barrels, the patches were wrapped in the opposite direction of the barrel twist. That is, they were wrapped so that as they went through the barrel, the rifling tended to unwrap them. In deeper grooved barrels—.003 inch and greater—the rifling slices the patch into small strips that are blown free at the muzzle. For use in these rifles, wrap the patch in the same direction as the barrel twist, so that the rifling tends to tighten the patch.

5. During the application of the patch, it is easy to overstretch the patch by pulling against the bullet as you roll it away from you. This results in a very tight patch, but often with overlapping ends. Once you get the feel for the correct amount of pressure, you can make every patch a perfect wrap and it will shrink tightly to the bullet.

8. In drying the patches, it is important to stand the bullets on their noses so that the patch can dry evenly and thoroughly, especially where it is tucked into the

cupped base or twisted into a tail. I like to set my loading block and bullets on the fireplace mantel or in the bright sun to dry. If I'm in a real hurry, I've been known to dry them in the oven. Set the oven to "Preheat" at 150 degrees F. and as soon as it is heated, turn it off and put the bullets in. By the time the oven has cooled, the patches are dry and shrunk tightly to the bullets.

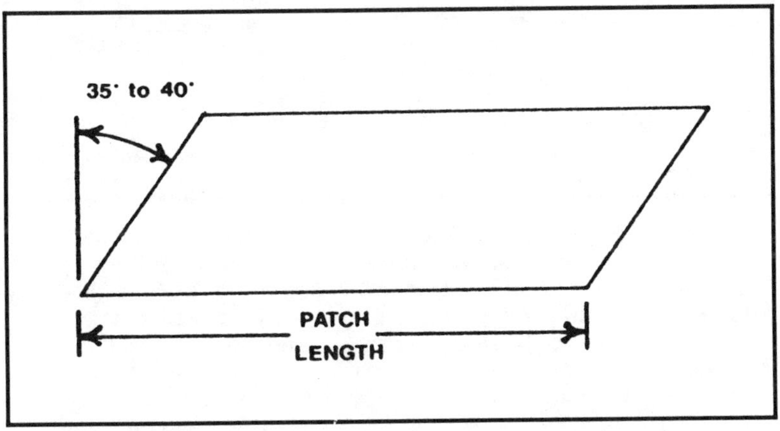

This is the way the patch should look before wrapping. Over the years, I've used patches cut at various angles up to 45 degrees, but I believe that a 35- or 40-degree angle is about optimum. The lesser angles are more difficult to start square, and the greater angles have more stretch in the narrow end of the patch.

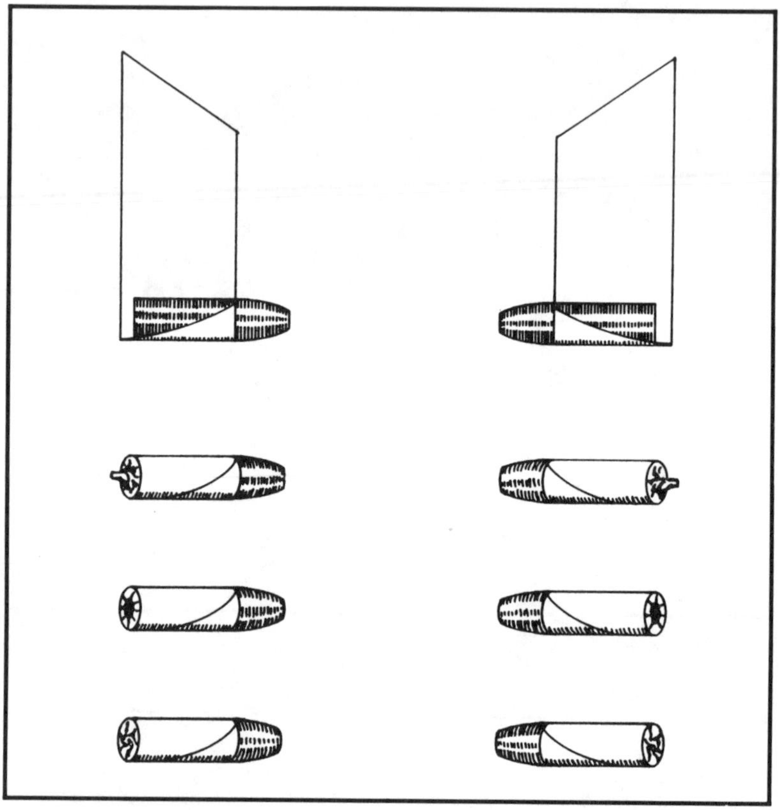

During the heyday of paper patched bullets, the direction of wrap was very important because the shallow rifling did not cut or slice the patch as is the case today. Because of that, patches were wrapped in the opposite direction of rifling twist so that the patch would be unwrapped as the bullet left the muzzle. Thus, the wrapping procedure shown at the right would be for a left-hand twist barrel, while the wrapping at the left is for a right-hand twist barrel. Whether the direction of wrap makes all that much difference today is debatable. The two upper bullets are examples of flat-based bullets and how the patch is cared for at the base. I prefer to twist the tail as shown on the upper bullet, and later trim it after it has dried. The bottom bullet shows the overhang of the patch pressed into the cavity of the cupped-base bullet.

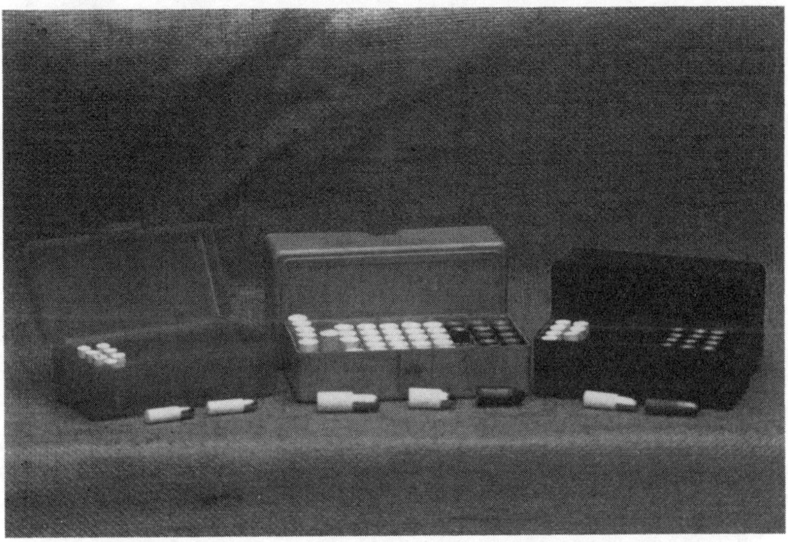

Paper patched bullets for .30-, .375- and .45-caliber rifles. I like to keep my bullets in slip-top plastic boxes intended primarily for pistol cartridges. This protects the bullets from dings and bruises.

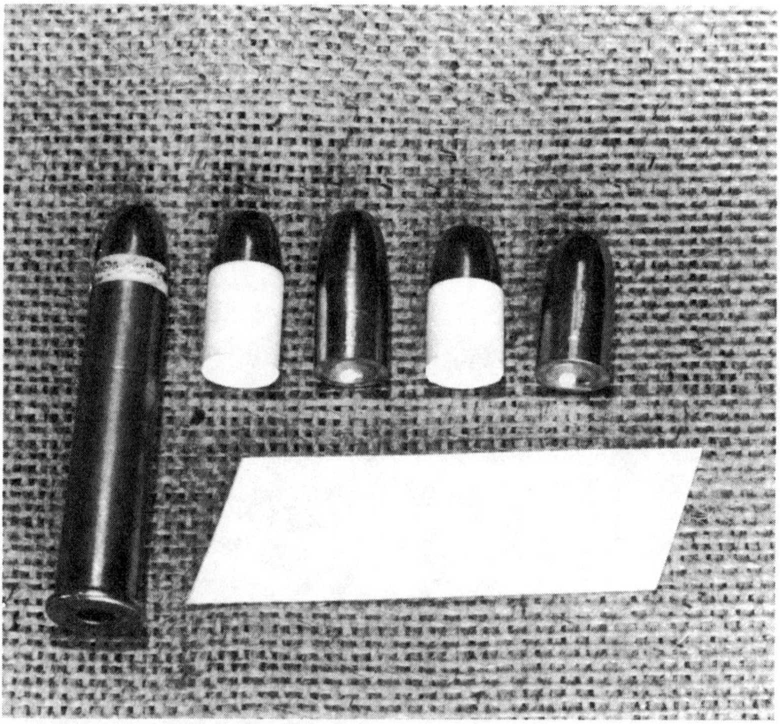

A .45-70 cartridge featuring a paper patched bullet, flanked on its right by patched and unpatched bullets weighing 408 and 363 grains, respectively. Below the bullets is a paper patch. These bullets and patch were made during my early days of paper patching bullets, and that is why the ends of the patch are cut at an angle far less than 40 degrees.

7

The Patching Board

Paper patched bullets are no different than any other bullet in the respect that uniformity begets uniformity. In other words, our goal is to make every bullet as identical as possible to its predecessor in weight, diameter, shape and patch application. Although the first three factors are taken care of in the bullet mould and/or swaging die, the final factor is quite vulnerable to human error.

Differences in patch location relative to the nose of the bullet, or oblique patches, can cause differences in point of impact—flyers. Even though patching a few hundred bullets makes one fairly adept at the process, the desire remains to more precisely locate the patch to ensure that it is rolled on square with the bullet's axis.

As far as I can determine, the women who worked in the factories applying patches to bullets, did so on a free-hand basis. There was no machinery to do such a job. However, I can't help but believe that with American ingenuity being what it is—and even more so in those days—that these women didn't devise some type of aid that helped patch more bullets per hour and better bullets in the process.

In an effort to achieve or even approach factory precision on paper patched bullets, I devised what I call a patching board. It might be a bit slower than freehand patching, but it is far more precise. With a little practice, each patch is rolled square with the bullet and every patch is located in the same relative position on the bullet.

To make a patching board, take a 6 × 9 × ¾-inch block of hard maple and make certain it is cut square across the ends. Then, on a table saw, cut a ⅛-inch deep groove just the width of the saw blade across one end of the block so that the outboard edge of the cut is about ⅛ inch from the block's end. Sand the block lightly and, using a red ball-point pen, scribe a three-inch line at right angles to the cut about two inches in from the edge. Then give the block a coat of polyurethane varnish.

To use the patching board, wet your patch and lay it on the board with the tail edge aligned with the red line and the end of the patch overhanging the end of the block. Lay your bullet on the patch over the ⅛-inch groove, thus ensuring it is square with the patch. Flip the end of the patch up over the bullet, pressing it against the bullet, and roll the bullet away from you. To accommodate different length bullets, make a reference mark to locate the nose of the bullet.

I find the patching board easy and fast to use. And once you get the hang of using it, patches are applied tighter than possible on a regular table top. This is probably because more pressure is put against the bullet to roll it out of the groove, and the pressure is maintained for the duration of the patching process. I press both forefingers against the patch and bullet when starting the roll to ensure that the leading edge of the patch is tight against the bullet.

The patching board works fine, as the narrow groove serves to square the bullet with the patch and requires little effort to roll the bullet out of the groove. Since .30-caliber bullets sink farther into the groove than the .375s and .45s, you must be careful about the amount of downward pressure on the bullet or you will tear the dampened patch.

I should mention that the coat of polyurethane varnish isn't necessarily for looks; its function is to prevent the wood from absorbing moisture from the patch. That happens fast enough as it is if you are patching bullets in the wintertime and have a good fire going in the kitchen stove!

Charlie Canoll sidestepped this problem by machining a patching board from a piece of aluminum. No absorption of moisture from the patch there!

Although I have made another type of patching board using glued-on pieces of $\frac{3}{8} \times \frac{1}{2} \times 4$-inch maple in place of the groove and the red-inked line, the grooved board is much faster and easier to work with. The wet patches stick to everything they touch and with the grooved board, if the patch isn't perfectly aligned, it is easy to lift off and realign. I believe that if we could turn the clock back 100 years, we would see young women and girls sitting at work benches on which they had scribed lines and reference marks with which to align their patches and bullet noses. They might even have had a groove in the bench to hold the bullet!

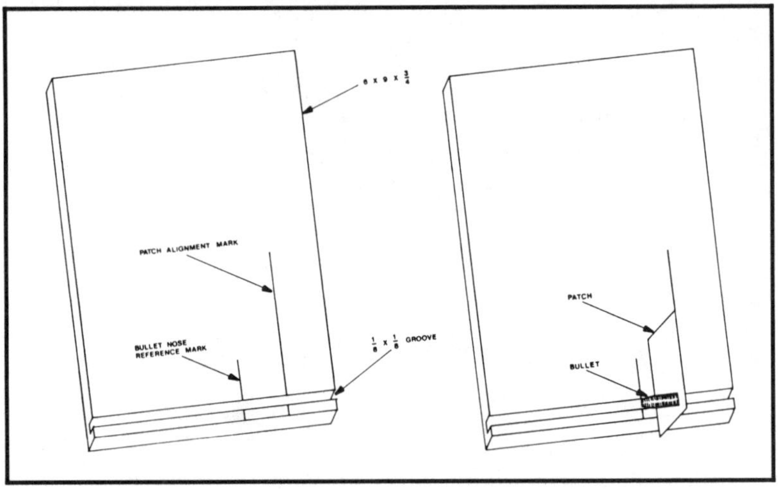

This is the patching board—easy to make, simple to use, and it does a good job helping those who roll their own to apply every patch identical to its predecessor. It is a great assist in ensuring that patches are consistently located in the proper spot on the bullet and that the bullet is square with the patch.

8

Lubrication

Paper patched bullets should be lubricated. Frank Mayer said he often used a bit of sperm oil, though his usual practice was to use a chamois skin soiled with graphite with which he wiped every bullet. He claimed this also helped waterproof his patches to some extent. At any rate, the lubricant was applied directly to the patch and was totally independent of any lube wads behind the bullet—which actually were used to soften the fouling.

If you are going to use smokeless loads in a modern rifle with a steep leade and sharp rifling, direct lubrication of the patch becomes an almost absolute necessity to prevent stripping of the patch in the bore. Since this is where I've used most of my paper patched bullets, I've had more than a few frustrating moments with patch lubes.

Over the years I have tried a number of lubricants, including the standard Alox/beeswax bullet lube, Rooster Laboratories' CFL-56 Case Forming Lube and their Minie Lube. I've tried Crisco™ and a mixture of Crisco™ and beeswax. And on top of all this, when I used the CFL-56, I also rolled the bullets in an oatmeal container in which I had placed a spoonful of fine graphite. This latter procedure worked fine, but is messy to handle.

The best patch lube I've used to date is a mixture of 45 percent (by weight) yellow beeswax with 55 percent clear Vaseline™. This is a clean, pleasant smelling lube that is easy to apply. It gives the patch some protection against moisture and it allows the cartridge to be chambered with the bullet kissing the rifling and then withdrawn with no damage to the patch. Not only do I use this lube on my paper patched bullets, I also use it on grooved bullets and on patches for my muzzleloader with hunting loads. It is an excellent all-around lube.

I lubricate my paper patched bullets just before seating by wiping a small amount of lube on the patch all around the bullet so that the patch is totally covered, except on the base. You need very little lube. A piece the size of a match head or slightly larger does a 400-grain .45-caliber bullet, so if you mix up a batch using 4½ ounces of beeswax and 5½ ounces of Vaseline™, you will have enough to last quite a while.

Although I have never experienced a problem of powder deterioration or contamination due to migration of the lube, the beeswax-Vaseline™ mixture will migrate to some extent. I tested this by putting a squashed-out lump of it on a small piece of paper and letting it stand overnight. Within 12 hours the migration stain was completely through the paper and had spread out to cover an area slightly larger than the lump of lubricant.

However, the amount of lubricant used in the test was several times the amount wiped on a bullet. Furthermore, when a paper patched bullet is seated in a cartridge case, much of the excess lube is squeezed out the case mouth onto the bullet. So, in practice, unless you are using a lube wad that contacts the bullet's base, there is very little lube left on the patch inside the case to cause a problem with migration. I have fired cartridges that had been loaded with smokeless powder for over two years and they still gave top performance with no indication of powder contamination.

If lube migration appears to be a threat, it can easily be countered in loads having 100 percent loading density, i.e., loads in which there is no air space between the powder and base of the bullet. Using an unsized empty case which has been trimmed square across the mouth and brought to a sharp edge using a Lee chamfering tool, cut some discs from a sheet of waxed paper and place one disc on top of the powder. The lube will not penetrate the waxed paper and the disc is so lightweight and thin that it will not affect the bullet's accuracy.

9

Waterproofing the Jacket

When I wrote *The Practical Paper Patched Bullet*, I included a short chapter on treating the patch with shellac and graphite to make it water repellant. After the book was published, I discovered that clear lacquer did an equally good, if not better, job, and unlike shellac it did not go bad in the can. I also wrote in the Introduction of that book that "we have whole generations of development ahead of us—waterproofing, durability and whatever else our imagination can envision."

Waterproofing the patch was a primary goal of mine. Once I had used the paper patched bullet for hunting and saw its potential, I forgot all about metal jacketed bullets. Metal jacketed bullets and grooved lubricated bullets, however, are not vulnerable to water—the paper patched bullet is. A single drop of water on the paper patch makes the bullet useless.

In his article "Making the Big-Bores Shoot" in the December 1934 issue of the *American Rifleman*, Frank Mayer wrote:

> When perfectly dry, slightly rub the forward end of the patch with a soft bit of chamois slightly soiled with pure graphite. This promotes true and easy passage of the bullet

through the barrel, and to some extent waterproofs the paper wrapping.

This simple statement, coming from a man who made his living with a rifle, tells me that during the era of which he wrote about there was no waterproofing of paper patched bullets within the United States. Certainly had such waterproofed bullets been available, the old-timers would have opted for them or devised a way to waterproof their own. A film of graphite on the paper patch has just about the same waterproofing characteristics as a patch without the graphite.

Despite this apparent vulnerability, for nearly 30 years most of the armies of Europe were armed with the paper patched bullet. They fought, fair weather or foul, under conditions that seemingly would have rendered such ammunition useless. Yet I have never read of that happening, nor have I read of any precautions being taken to prevent its happening. It is my opinion that waterproofing of paper patched bullets was so common in England and Europe, at least for those bullets used by the military, that it was too insignificant to warrant mention and was not even recognized for what it was in the United States.

This came to light when Fred Cornell made a simple remark to me one evening. "I have a friend," he said, "who has a pail full of old 11.15×58R Austrian Werndl cartridges loaded with paper patched bullets. And those bullets have been lubricated by dipping them in beeswax."

Then Charlie Canoll told me he had some .41 Swiss Vetterli cartridges in their original box and they had beeswax on the patch. Sure enough, they did. More than that, the Vetterli ammunition appeared to have been patched *after* the bullet was seated, and the patch applied so as to cover the neck of the case. This being true, then it follows that the beeswax was applied to the bullet and patch after the cartridge was loaded.

Much later, two articles appeared (one in *Rifle* and one in *Handloader*) that contributed further evidence of the widespread use of beeswax on paper patched bullets. In *Rifle* No. 116 on page 32, author Dennis Bruns (in writing of the .577/450) made the simple statement: "The bullet was paper patched and covered with beeswax."

Gun writer Gil Sengel greatly elaborated on this same bullet on page 9 of *Handloader* No. 145. He wrote:

> The British did not favor a grooved, lubricated bullet. Instead, a paper patched slug made from an alloy of twelve parts lead and one of tin was used. Patch wrapping in the direction of rotation caused the paper to fall from the bullet as it exited the muzzle. Dipping the completed bullet in melted beeswax provided patch lubrication. After cooling, pushing the slug through a heated die removed excess wax. Optimum diameter was .458 inch although bullets in commercial ammo range from .455 to .470 inch.

Now as a lubricant, beeswax by itself is pretty poor. Its primary use in bullet lubes is as a vehicle by which the lube is carried and tempered to hardness. In short, it is my opinion that the primary function of the beeswax on the paper patched bullets of the .577/450, .41 Swiss Vetterli and the Austrian Werndl cartridges was for waterproofing purposes and not lubrication. Lubrication may have been a secondary function as was an increase in patch durability, but the real purpose of the beeswax was to waterproof the patch so that the ammunition could be used during the torrential downpours of the tropics and so that the patches would not be affected by high humidity. I want to emphasize that this is an opinion, not a historical fact.

With information on the .41 Swiss Vetterli and someplace in the back of my mind a picture of .22 rimfire cartridges suspended over a tray of melted lubricant, I set about waterproofing paper patched bullets *after* they had been seated in loaded .45-70 cases. In essence, I suspended loaded .45-70 cartridges over a tray of melted beeswax in such a fashion that the entire bullet and about $\frac{1}{16}$-inch of the case were submerged for three minutes. Afterward, I took them out, shook off the excess wax and stood the cartridges bullet down in a glass of water for 20 minutes.

Then I took them out and shot them over the bench. Waterproofing was a reality!

Although I believe the procedure I used was safe enough, I didn't like the idea of suspending loaded cartridges in melted beeswax, nor do I recommend it. The cartridge case quickly heats up and possibly could "cook off." Thus, instead of waterproofing loaded cartridges, I set about waterproofing the bullets before they were seated in the cartridge case. This is the simplest, best and safest way to do the job.

Equipment for the waterproofing process is easy to prepare. First, you need a double boiler consisting of two bread pans, one being 4½ × 8½ inches on the inside, and the other being 5¼ × 9¼ inches on the inside. Then you need a piece of ⅛-inch thick Plexiglas™ or stiff piece of sheet metal about 6 × 10¼ inches. Cut a 4 × 8-inch opening in this with the corners of the opening rounded to a ⅝-inch radius.

In operation, fill the large bread pan three-quarters full of water, set the Plexiglas™ collar on it and then place the smaller bread pan containing the beeswax in the collar's opening.

Using a piece of ¼-inch hardware cloth, make a small tray 3¼ × 6¼ inches with turned-up sides and stiff wire handles made from a coat hanger at each end.

One of the main problems encountered in waterproofing is leaving too much beeswax on the bullet as this can interfere with chambering a loaded cartridge. This is why the British pushed their .577/450 bullets through a heated die. Yellow beeswax melts at 146.4 °F. and must be much hotter to run freely off the bullet. Further, the bullet must be heated all the way through so that it doesn't cool the wax prematurely, thus preventing a free runoff.

To accomplish all of this, I use the following techniques when waterproofing paper patched bullets.

1. Keep the water in the double boiler at a low boil.

2. Let the beeswax heat for a full five minutes *after* it has melted and before immersing any bullets in it.

3. Lay 15 or 20 bullets in the wire mesh tray and submerge them in the melted beeswax for 10 minutes.

4. After the bullets are thoroughly cooked, lift the tray from the double boiler, shaking it gently to let the excess wax drip off, and then gently roll the bullets out of the tray into a third bread pan in which a paper towel has been folded to cover the bottom of the pan.

5. Roll the bullets back and forth on the paper towel by tilting the pan, until all of the excess wax has been absorbed by the toweling and the bullets have cooled to the extent that they do not stick to the toweling.

6. After the wax is thoroughly cooled and set, wipe some finely powdered graphite on each bullet.

7. Let the bullets stand for a day before loading and shooting. This is to let the beeswax shrink to its maximum and thus not stick to the bullet.

This last item is quite important. After I had tried the beeswax and found out how well it worked, I melted up a pan full of layout wax—a wax used to paste down typeset galleys—and waterproofed a dozen bullets with it. Whereas yellow beeswax has a high shrinkage factor and doesn't stick to the bullet, the layout wax had practically no shrinkage, and pieces of the patch rode the bullet all the way to the target. More than that, in some cases bits of the patch stuck to the inside of the cartridge case, thus damaging the patch before it ever went down the barrel. Accuracy was badly affected.

Actually, I do not waterproof all of my bullets, but only enough to have on hand for hunting season. As far as accuracy is concerned, properly waterproofed bullets shoot just as well as those that are not treated.

An early attempt at waterproofing paper patched bullets was made by standing some bullets in a jar lid containing Alox/beeswax bullet lube, and putting the whole works in an oven at 150° F. The lube melted and soaked up into the patch. Had I used yellow beeswax, it might have worked, but the double boiler method is far better and safer.

The double boiler made from two bread pans and a sheet of ⅛-inch Plexiglas™.

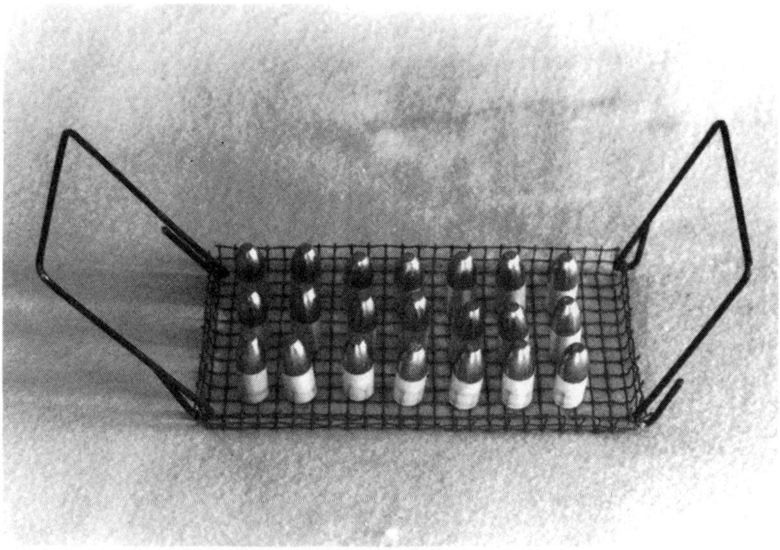

The wire mesh tray made of ¼-inch hardware cloth and a coat hanger. Here the bullets are shown standing up. Since they will usually fall over anyway, I now lay them down.

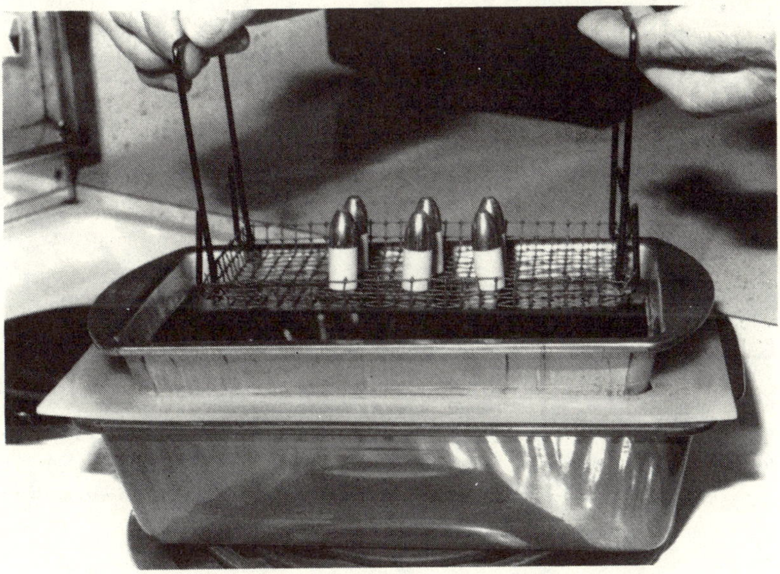

Loading the tray of bullets into the double boiler of melted beeswax.

Some .45-caliber paper patched bullets. Those on the left—the darker ones—have been waterproofed and rubbed with graphite.

10

Cutting Patches in Quantity

Cutting a quantity of patches can be a tedious job, especially if you use the old time-honored method of a metal template and a razor blade. The best method I have found is to first obtain a good steel ruler 12 inches long and $^{15}/_{16}$ or one inch wide. Then, take your sheet of patching paper and, using the ruler as a straightedge and a utility knife with a sharp blade for cutting, cut the sheet into strips of the proper width for your patches.

On one strip, measure out the length of as many patches as that strip will accommodate and, using a plastic triangle and sharp pencil, mark out the angled ends of each patch. Now, using this strip as your pattern, back it up with two or three blank strips, tapping them against the table top to get them all even. Once they are even, hold them firmly and, with a pair of sharp scissors, cut through all of the strips following the angled pencil lines on your pattern.

Using this method, you can cut a lot of patches in a relatively short time.

If you really go whole hog for paper patched bullets, you will reach the point where you need to be able to cut a perfect patch every time and do it in such quantities that one session of patch cutting will cover several sessions of casting and patching. Although scissors and razor blades work well for the low-volume shooter, their output is disappointing and tedious for the high-volume shooter. True, the rectangular patch lends itself fairly well to high volume with a pair of scissors, but high-volume precision cutting of the angled patch is a horse of a different color.

The shooter who goes high volume with paper patched bullets and who demands precision-cut traditional patches, needs a good, high-quality paper cutter modified by the addition of a paper patch cutting guide. Yes, such a cutter will cost a few shekels, but when you consider how few metal jacketed bullets you can buy for the same price, a good paper cutter is a bargain.

For my own personal use, I purchased a 15-inch Premier No. 215 Trimmer with a solid wood base, heavy steel blade, safety guard and steel rule. Premier also offers a No. 715 Stack Cut All-Metal Trimmer with a hold-down on which you can stack up to ¼-inch of paper.

To make a paper patch cutting guide, enlist the help of someone with woodworking equipment, and have them make a guide from a piece of dry hardwood such as hard maple, oak or cherry. A short piece of oak flooring is fine. Cut the guide to the dimensions shown in the illustration at the end of this chapter, making certain all edges and surfaces are smooth and straight. The angle cut at the end of the guide determines the angle on the end of the paper patch. The guide shown is for cutting a 40-degree angled patch. If you prefer a different angle on the patch, say 30 degrees, then cut the angle on the guide at 60 degrees.

In order to mount the guide on the trimmer, place the angled end of the guide about ⅛-inch back and parallel to the cutting edge of the trimmer. Drill three or four evenly spaced holes down through the guide and trimmer to accommodate 5/16-inch diameter bolts 1½ inches long. Run the bolts down through from the top, using a flat washer under the head and a second flat washer under the wing nut at the bottom.

Once you have the guide secured in place, then you can measure off the length of your patches, beginning at the cutting edge and working back along the guide. Place a fine mark on the guide for the length of each patch and indicate its caliber—.45, .44, .40, etc. If desired, you can mark out the length of more than one caliber on the same guide, or make a guide for each caliber.

If you are reluctant to put permanent patch-length reference marks on the guide, you can place a strip of 5/6-inch correction tape along the sloping edge of the guide on which to mark out the length of the patches. That allows removal and replacement of the tape.

Obviously, when using the trimmer to cut strips, the guide must be removed. I usually take three to six sheets of patching paper and, after aligning them, staple them together in one corner. Then, while pressing against the entire stack with a stiff ruler held close to the cutting edge, I cut off the strips, setting each stack of strips off by itself.

Once all the strips are cut, I carefully align each stack and staple it in one corner. Then with the cutting guide bolted in place, I proceed to cut the stacks of strips into individual patches. Actually, it takes longer to remove and replace the cutting guide than to do the actual cutting. Normally I stack five sheets at a time and cut about 15 sheets at a sitting. This gives me 330 patches for .45-caliber bullets.

It might be worth mentioning that with my patches cut at a 40-degree angle, I get 22 patches per sheet, whereas if I make the patches rectangularly shaped, I get 33 patches per sheet. Thus, there is a lot of waste by cutting at a 40-degree angle. Perhaps if the patches were cut at a lesser angle, say 30 or even 25 degrees, one could get 33 patches per sheet with no waste.

I personally don't care for the lesser-angled patches. I believe the 40-degree angle results in a tighter patch which is easier to align and to apply, but the option of lesser angles is mentioned here for those whose paper availability might dictate the extreme in economy.

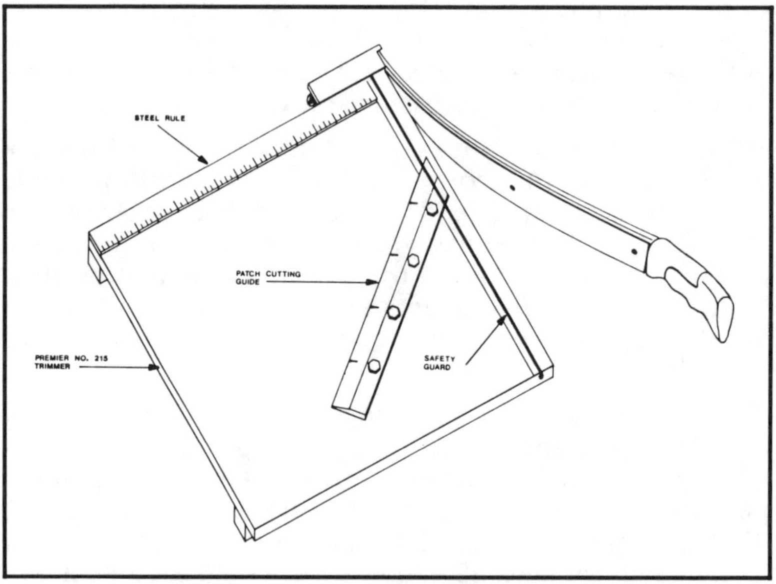

A Premier No. 215 Trimmer modified by adding a paper patch cutting guide. With an outfit like this, the paper patched bullet shooter can cut enough perfect patches in one hour to last many shooting sessions. The 15-inch trimmer is strongly recommended over the 12-inch because it is far heavier and has a more rugged blade. When using this board, work the blade smartly—don't hesitate—*and keep your fingers away from the blade*!

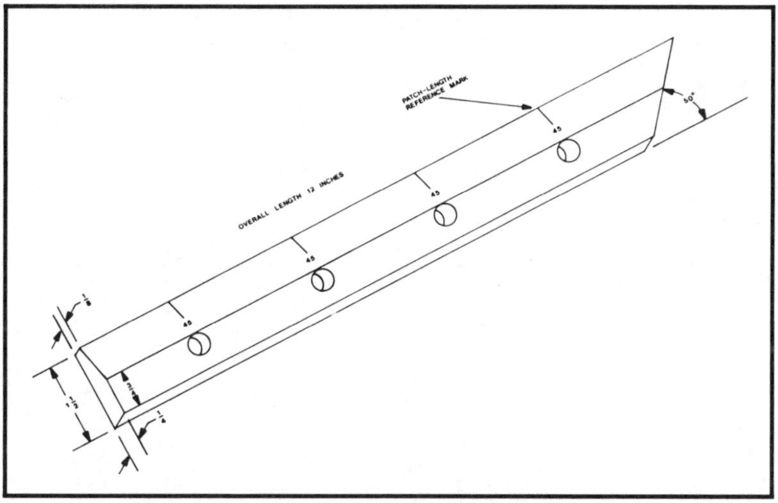

The paper patch cutting guide used with the Premier No. 215 Trimmer. Although this shows the mounting holes and patch length reference marks, the reference marks should not be put on until after the guide is mounted on the trimmer. The holes should be drilled through the guide and trimmer at the same time to ensure positive alignment.

11

Loading the Paper Patched Bullet

Paper patched bullets are most vulnerable during the loading process, especially when seating the bullet in the case. Unlike metal jacketed bullets which are seated as though they are a neck expanding plug, or even the grooved lubricated bullet which sometimes receives the more gentle treatment of a Lyman M-die, the paper patched bullet must be handled with kid gloves during the bullet seating process.

Although old-timers often were able to hand-seat their bullets friction tight in the unsized cartridge case, most of today's chambers are so large that the cartridge case must be resized, or at least neck sized, at each loading. If black powder or Pyrodex is the propellant, neck sizing alone usually will suffice and is preferable since it helps extend case life as well as case capacity. If smokeless powders are used, especially for heavy loads, then full-length-resizing is usually in order to avoid a "sticky" cartridge case in the chamber. In either event, the neck of the case is almost always reduced at each loading and must be opened up to accept the paper patched bullet.

The thicker the patch is, the more resilient it is and the less likely it is to tear. As the *unpatched* bullet's diameter increases and the patch's thickness decreases, however, the patch becomes increasingly vulnerable during the seating operation.

With thick patches and under-bore-size bullets, the Lyman M-die does a passable job, running the case onto the plug just as one would for a grooved bullet. In other words, the die uses both step diameters on the plug, but does not bell the case mouth.

When using bore-size or over-bore-size bullets and thin patches, we are faced with the problem of belling the case mouth by running it a bit farther up on the M-die. However functional this might be, using the M-die for belling the case greatly overworks the brass and soon results in cracks around the case mouth.

Cases for use with paper patched bullets should be *flared* at a gentle angle just enough to accept the patched bullet. This requires a neck expanding plug that has a slight taper to it.

For use in preparing .45-70 cases, I had the neck plug from my RCBS plug die modified as shown in the first drawing at the end of this chapter. It works perfectly and causes minimum stress to the brass. This tapered neck plug worked so well for my .45-70 brass, that I soon asked Charlie Canoll to make similar neck plugs for .30-caliber and .375-caliber. Charlie did this, maintaining all of the dimensions shown on the drawing except for the taper and the pilot section. The pilot section was machined to .001 inch under the jacketed bullet diameter for each caliber, and the large end of the taper was .015-inch larger than the jacketed bullet diameter.

The tapered neck plug opens up the case mouth with such a gentle flare that it is barely discernible. It causes little stressing of the brass and can be adjusted to accept as much or little of the bullet shank as desired. Bullets are seated with so little pressure there is absolutely no distortion of either the nose or the base, yet they are held firmly in the case so that they can be carried afield or in a cartridge belt with no danger of the bullet falling out.

Today, I not only use the tapered neck plug for my paper patched bullets, but also use it whenever seating grooved lubricated bullets. *It is the one accessory that no cast bullet shooter—paper patched or grooved lubricated—should be without.*

When we get into the very basics of reloading the paper patched bullet, there is one factor that stands out above all others: The paper patched bullet does its best under 100 percent loading density situations. Although I believe this is true with all bullets, I propose it is more apparent with the paper patched bullet because of the lower pressures involved. It has always been my observation that with a given charge of powder behind identical bullet weights, metal jacketed bullets deliver the highest pressures and paper patched bullets give the lowest; lubricated bullets fall somewhere in between. I hasten to emphasize that this is an observation on my part and not a proven fact. As you approach full-throttle loads, the differences in apparent pressure lessen appreciably.

What it all adds up to, at least for me, is that I seldom load a paper patched bullet leaving an air space between the powder and bullet. In some cases this may dictate a slower powder than normally used and in other cases it may require a Dacron or Cream of Wheat filler. Although I use fillers in many of my loads, I'm not going to recommend them because of the fact that so many shooters have put a ring in the chamber of one of their pet rifles through the use of a filler. If you insist on using a filler as I do, make absolutely certain there is no air space between the filler and the base of the bullet. To leave an air space is to make the filler respond as a potential projectile and the bullet as an obstruction.

Slower than normal powders loaded to 100 percent loading density sometimes give exceedingly good results. In loading for his Remington Model 788 .30-30, Charlie Canoll uses 32 grains of IMR-4320 behind a 175-grain smooth-sided pure-lead paper patched bullet for an average of 2,198 pfs velocity. Now that's a shade better then factory ballistics with all the accuracy a man could hope for. The load is pure murder on the woodchucks that seem insistent on getting into Charlie's garden.

Case capacity of the .30-30 is just about perfect for paper patched bullets in the 170- to180-grain weight range. As indicated with Charlie's load, a 100 percent loading density with no filler required gives near-maximum velocity for a pure-lead paper patched bullet from a 10-inch twist barrel.

The one thing which is remarkable about the soft paper patched bullet is its compatibility with fast-twist barrels. Chapter 12 further explains why softer bullets more tightly grip the rifling. The very fact that Charlie is driving his bullets at jacketed bullet velocity is strong evidence of this phenomenon.

A few years ago my nephew, Joe Piccolo, had a wildcat rifle made up on a Ruger No. 3 action. This was what he called the .37 Rimmed, a .30-40 case blown out straight and necked down to .375 caliber. Actually it was just about the optimum size case for 100 percent loading density charges for paper patched bullets of .375 caliber.

One particular load we used involved a 299-grain paper patched bullet backed by 46 grains of IMR-3031. This load gave an average velocity of 2,180 fps with an extreme spread of 26 fps for five shots. Since the unpatched diameter of the bullet used was .363 inch, and should have been about .367 or .368 inch, I wasn't surprised that accuracy wasn't up to par.

Then in a moment of experimentation, I boosted the charge to 48 grains of IMR-3031—a full 100 percent loading density—and reduced the group's size to less than two inches with a substantial increase in velocity to boot!

Now this load is superior to the old .405 Winchester, which was certainly no slouch. Although bullet weight and velocity are nearly identical, the sectional density and ballistic coefficient of the .375 paper patched bullet is far superior to the .405 Winchester's 300-grain bullet.

In loading paper patched bullets for my Ruger No. 1 .375 H&H, I used 47 grains of IMR-3031 and filled the rest of the case with Cream of Wheat which was slightly compressed when the bullet was seated. In this case I used a 310-grain soft bullet—the mould had been corrected to give an unpatched diameter of .371 at the base tapering to .367 at the ogive—for an average velocity of 1,982 fps with an extreme spread of only 18 fps. Accuracy was superb. I also tried this same bullet over 82 grains of H-4831, but was never able to get any kind of decent accuracy with it.

I must admit that my first love in cartridges is the .45-70, and it is with this cartridge that I have loaded and fired thousands of paper patched bullets of different diameters, styles, alloys and propellants. The .45-70 is an extremely versatile cartridge and I rate it as the best cast bullet hunting cartridge in existence today, bar none! Although I have used dozens of various cast bullets in several different .45-70s, it is the paper patched bullet I like the best, especially in the single shots.

From the date that I first acquired a Winchester 1886 .45-70 capable of handling the more powerful loads, my standard hunting load was the one so much admired and promoted by Elmer Keith—53 grains of IMR-3031 with a 400-grain bullet. I have used this load off and on for 40 years in several different rifles, and it has always proved excellent. In my strong Ruger single shots, I have used the same charge with bullets as heavy as 425 grains, and in the same rifles I have used much lighter bullets with much heavier charges. For example, with a 355-grain paper patched bullet, I increased the charge to 55 grains, and with a 300-grain RCBS bullet cast of Linotype, I increased the charge to 60 grains.

The important fact here is that every load I used had 100 percent loading density—the bullet was seated on top of the powder. When used in the straight-walled .45-70 case with the bullet weights mentioned, IMR-3031 powder is just about as slow a burning powder as can be used and still produce uniform ignition. Even with the 400-grain bullet and 53 grains of IMR-3031, there are usually traces of unburned powder left in the barrel.

In chronographing the 53-grain load with a 405-grain paper patched bullet over the Oehler 33D Chronotach, I got an average velocity of 1,833 fps with an extreme spread of 25 fps. That is a shade better than 3,000 foot-pounds of muzzle energy! Not bad for a piece of lead pipe wrapped in a scrap of paper!

For heavier bullets—those in excess of 425 grains—and full-throttle loads, I use the slower burning IMR-4064 powder, again loaded to 100 percent loading density. As mentioned earlier, 60 grains of this powder under a 475-grain Postell bullet patched with nine-pound onionskin gave an average velocity of 1,772 fps with an extreme spread of 52 fps.

Working toward the other end of the scale for moderate velocity, easy shooting loads, I find that SR-4759 powder delivers outstanding performance, especially if the air space is filled with Dacron. In my Navy Arms rolling block, the one load that will repeatedly shoot just as closely as I can hold is a 480-grain swaged paper patched bullet backed by 27 grains of SR-4759 and a piece of Dacron batting ½ × ½ × ¾-inch long.

Now over the years I've been led to believe that paper patched bullets didn't function too well in the lever action repeating rifles because of the patch's vulnerability during its journey through the loading gate and feeding mechanism. Where I got this notion, I don't remember, but it is wrong. Loaded properly, the paper patched bullet will feed and function just as well in a lever action rifle as in a single shot.

There are two ways to protect the patch for use in a lever action rifle. First, waterproof the patch as described in Chapter 9. The beeswax and graphite on the patch do a lot toward promoting durability. Secondly, and even better than the waterproofing, is to regulate the width of the patch so that when the bullet is seated to provide the required overall length of the cartridge, the entire patch is enclosed within the cartridge case. I have used both of these methods in the new Marlin 1895SS and although I much prefer the latter method, either way will work fine.

As stated earlier, my Marlin 1895SS .45-70 has one of the new deep-grooved Micro Groove barrels. The Marlin handles all types of cast bullets, paper patched or grooved lubricated, with full-throttle loads and alloys as soft as pure lead with 3 percent tin. It is perhaps the finest cast bullet hunting rifle presently on the market.

However, having 12 lands and 12 deep grooves, these new Micro Groove barrels bring to light another aspect of paper patched bullets not usually heard about—displacement of patching material.

Since the bore diameter of my Marlin barrel was a bit smaller than that of my Rugers, I used a bullet having an unpatched diameter of .450 inch, and I patched it with .003-inch thick 16-pound paper instead of the usual .0025-inch thick nine-pound onionskin. These bullets worked fine and shot well. When I tried to use a .452 inch diameter bullet patched with onionskin, however, accuracy deteriorated and I picked up some leading in the barrel.

I have to believe that the deep-grooved Micro Groove barrel having 12 lands and grooves displaces so much of the paper patch that the thinner nine-pound onionskin lacked the necessary resilience and exposed some of the bore to the naked bullet.

Mention of loading for lever action rifles brings up another point, or perhaps explodes another myth about paper patched bullets. Many times we have been told, and I've said it myself, that you can't crimp a paper patched bullet. That is wrong. You can (and I have) crimp paper patched bullets, although with more caution and care than is usually exercised with the grooved lubricated bullet.

Flaring the mouth of the cartridge case to accept the paper patched bullet offers no problem for most rifles. In fact, after the bullet is seated, evidence that the case was ever flared is almost imperceptible. Although cartridges with flared cases will feed through the Marlin with no hang-ups, you can sometimes feel the lip of the cartridge case momentarily snag the lip of the chamber when the breech bolt is levered forward.

I like my cartridges to feed smoothly without a hitch. So whenever I load paper patched bullets for the Marlin 1895SS, after the bullets are seated to the proper depth, I remove the decapping pin assembly from the full-length resizing die and run the loaded cartridge into the die just far enough to remove any flare from the mouth of the case. In effect, this is a tapered crimp, but it works just fine with no ill effects on the paper patch.

This, of course, is not what we normally think of as crimping. In the November 1988 issue of *Guns* magazine, there appeared an article by Raymond Page in which he told of using paper patched bullets in .44 Magnum and .45 Colt revolvers and Winchester carbines. These were full loads and the bullet was severely crimped to keep it from shifting under recoil. In this article Page wrote:

> The paper wrap below the crimp line must remain undamaged during the bullet seating process. Don't be concerned if your crimp pushes the case mouth into the paper. As long as the paper wrapping below the crimp line remains intact, this will provide the necessary protection to the bullet.

Although I am not a handgunner, and semiwadcutter bullets are not the best design for paper patching. I did patch some of my RCBS 44-240-SWC bullets, crimped them and fired them through my Marlin 1894 Sporter .44 Magnum. They worked without a hitch.

Returning to the few lines quoted from Page's article, you will note that he stresses the importance of an intact paper patch. Nothing could be more vital to the performance of the paper patched bullet, and nothing can contribute more to the security of the paper patch than the tapered neck plug mentioned at the beginning of this chapter.

Now, thus far we have done a lot of talking about loading paper patched bullets on top of smokeless powder. But don't let that fool you into thinking that black powder or Pyrodex is some archaic form of propellant best relegated to the dim distant past or to a group of those who like to play at being a part of that dim distant past. Far from it, properly assembled loads of soft paper patched bullets over black powder or Pyrodex provide superb accuracy with excellent killing power and relatively low recoil. In fact, if I were using a rifle of black powder vintage or a replica of a black powder cartridge rifle, unless I knew for a certainty that the action would handle smokeless loads, I would stick to black powder or Pyrodex with soft paper patched bullets and not feel the least bit undergunned.

Looking at the practical side of black powder or Pyrodex, we find that for best performance both propellants are *always* loaded to 100 percent loading density. More than that, both must be compacted within the cartridge case for best ignition. In the case of black powder, this means dribbling the powder down through a drop tube that can be anywhere from 12 to 36 inches. The powder will arrange itself within the case so as to take up a minimum of room without fracturing or crushing any of the kernels, as usually happens when black powder is mechanically compacted during the bullet seating operation.

On the other hand, Pyrodex does not settle itself in the case to the same extent as black powder. In the .45-70 case I usually compress Pyrodex to a specified level by using an empty .35 Remington case held against a decapping rod from which the decapping pin has been removed. An inked line on the case registers the depth I want and a single stroke on the loading press does the rest.

Another practical aspect of black powder or Pyrodex is that just as long as you use the correct granulation, there is no way you can put an excessive charge in the case and still seat a bullet. This means that with black powder you use Fg or FFg for most rifle cartridges, and FFFg for handgun cartridges regardless of whether they are used in a rifle or handgun.

Although there is a trend among muzzleloaders to use FFFg in just about any bore size because of its cleaner burning, Major Don Holmes, in writing of the 1,000-yard Canadian black powder matches, stated in *Rifle* No. 100:

> Gearhart-Owen GOEX Fg black powder gives very uniform results. FFg may also be used. The most consistently accurate ammunition has the powder weighed, drop-tubed into the case and never crushed during bullet seating.

The shooter using Pyrodex should use CTG or RS grades for rifle cartridges, and grade P for handgun cartridges.

Finally, by using black powder or Pyrodex, the reloader can easily get by without the use of a powder measure or powder scales, although I personally prefer to weigh all of my charges.

Both black powder and Pyrodex foul a barrel, and black powder fouling accumulates to the point of zero accuracy. With Pyrodex, however, you have as much fouling after the first shot as you will have after the tenth. In either case, unless something is done to reduce the fouling or to keep it very soft, accuracy will leave a lot to be desired unless the barrel is cleaned after every shot.

Although cleaning after every shot may be just fine on the target range, and a necessity for muzzleloading the paper patched bullet, it is highly impractical in the hunting rifle. Here we want to be able to fire a string of shots and know that the accuracy of each bullet is unaffected by its predecessor.

There are two ways to accomplish this objective. The first, and I believe the most practical, is to use a duplex load involving a priming charge of smokeless, preferably

SR-4759, and a main charge of black. This helps burn off the solids, thus keeping the barrel relatively clean from shot to shot.

When I got my Ruger No. 1 .458, the first thing I discovered was that I could dribble 90 grains of FFg in this case and still have room to seat a 550-grain paper patched bullet. Armed with this rifle and a handful of these black powder cartridges, I went to the range and sent the first bullet off amid a cloud of fire and smoke. It was beautiful! That is until I found that I couldn't chamber a second cartridge due to the fouling in the throat.

Back home again I reloaded those cartridges using 12 grains of SR-4759 as the priming charge and 72 grains of FFg as the main charge, and from that point on I could shoot as long as I wanted with close to two-inch accuracy with those long paper patched bullets.

When using duplex loads in black powder vintage rifles, we have to remember that the smokeless priming charge does increase pressures, and probably shouldn't exceed 10 percent of the rated black powder charge for that cartridge. Thus, for a .45-70 of black powder vintage, the priming charge should not exceed seven grains and probably would be better if kept to five. For strong-action rifles like the Ruger single shots, it has been my experience that a 13 percent priming charge of SR-4759 delivers about the best accuracy.

One more thing to note about duplex loads is that it is well recognized among serious black powder shooters that one grain of SR-4759 is the equivalent of three grains of black. Thus, in my load for the .458, although I reduced the charge of black from 90 to 72 grains (a reduction of 18 grains) the 12-grain priming charge resulted in an equivalent *increase* of 36 grains!

The purist who insists on using straight black powder or Pyrodex with paper patched bullets must do two things in order to ensure easy chambering of every cartridge and good accuracy every time he presses the trigger. First, he must use a soft alloy bullet having a patched diameter of .001 or .002 inch larger than the rifle's bore diameter. Secondly, because the paper patched bullet carries very little lubricant with it, the shooter must load his ammunition with some type of a grease wad that keeps the fouling soft enough to be removed on a single pass with a snug-fitting dry patch.

As mentioned elsewhere in this book, the first requirement can be accomplished in two different ways: either by casting a bullet of the proper size to begin with, or by running a patched bullet through a reducing die to bring it down to the proper size. If you elect to use a mould, it should cast a bullet .004 to .006 inch smaller than your rifle's bore diameter. If you use a reducing die such as the ones produced by Dave Corbin of White City, Oregon, get a die that's equal to the bore diameter of the rifle or .0005 inch over.

Since I already had a Corbin reducing die for reducing the diameter of my lubricated bullets from .459 or .460 inch to .452 inch, that is the die I use. A soft paper patched bullet measuring .459 inch over the patch measures .453 inch after it has been pushed through the die. This system works to perfection, but I must emphasize the superiority of the Corbin reducing die as compared to a standard sizer-lubricator die.

Accomplishing the second requirement took a bit of experimenting. Although the usual practice handed down to us over the years is to use a thin card wad on top of the powder followed by the grease wad, I have never found this to be very functional. In fact, with cupped-base bullets, many times I have found both wads sledged into the bullet's base with very little effect on the fouling in the barrel and considerable effect on accuracy. Many other times I have picked up both the grease cookie and card wad stuck together at random points between the rifle's muzzle and the target.

It is my opinion—others may see it differently—that for a grease wad to be effective in keeping the fouling soft, it must vaporize in the barrel. It cannot do this with a thin card wad between it and the powder gases. In short, the card wad must disintegrate or vaporize along with the grease wad.

To accomplish this, I use a disc of waxed paper cut to size by sharpening the mouth of a fired case with a Lee chamfering tool. One or two taps with a four-ounce plastic hammer on the head of the cartridge case cuts the discs cleanly and quickly. For my grease wad, I roll up a ball of my patch lubricant about the size of a .45-caliber roundball between the palms of my hands and place it in the case on top of the waxed paper disc. Then I seat my paper patched bullet and I'm in business—almost.

These .45-70 cartridges will stay under two inches at 100 yards—usually well under. The fouling they leave in the barrel is so soft that one pass with a snug fitting dry patch will remove virtually all of it.

If such cartridges are to be loaded for any length of time before shooting—say longer than two weeks—then something must be done to prevent migration of the lubricant into the patch. I found this out by loading several cartridges and then letting them stand for 2½ months. During that time the lubricant permeated the patches to the extent that they were translucent and were no longer sliced properly by the rifling. Groups went from well under two inches to four and five inches. I also broke down some of this ammunition to check the powder, and from what I could determine there had been no migration of the lubricant past the disc of waxed paper.

The most positive way to prevent deterioration of the patch due to lube migration is to waterproof the bullets as described earlier. There is no way the lubricant can penetrate the wax in the patch any more than it can penetrate the disc of waxed paper between itself and the powder. Another way to prevent migration, and perhaps the simplest, is to place a second disc of waxed paper on top of the lubricant to prevent contact with the bullet's base.

It is interesting to note that although a quantity of lubricant behind the bullet will ruin the patch within a few weeks, the amount of lube normally applied to the patch for lubricating purposes does no damage whatsoever. I have fired paper patched ammunition that has been loaded for over two years without a hint of a problem or any loss of accuracy.

One final note. If you are using Pyrodex with undersize paper patched bullets, and depend upon the force of the Pyrodex to expand the bullet to grip the rifling, my suggestion is to use RS Pyrodex instead of CTG Pyrodex. This latter grade will not bump a bullet up on the barrel—at least that was my experience when using 80 grains of it behind a muzzleloaded 550-grain paper patched bullet in a Navy Arms rolling block. With groove diameter bullets, however, the CTG Pyrodex works just fine.

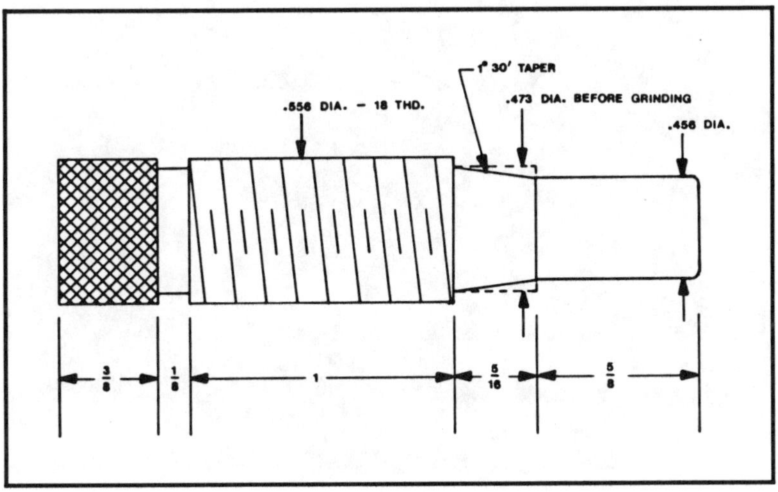

Dimensions of the neck plug from an RCBS .45-70 plug die modified to flare the inside of the case neck. The tapered section shown here is greatly exaggerated for clarification. Using a tapered plug to flare the case mouth works the brass at a minimum and ensures that the patch is not torn when the bullet is seated.

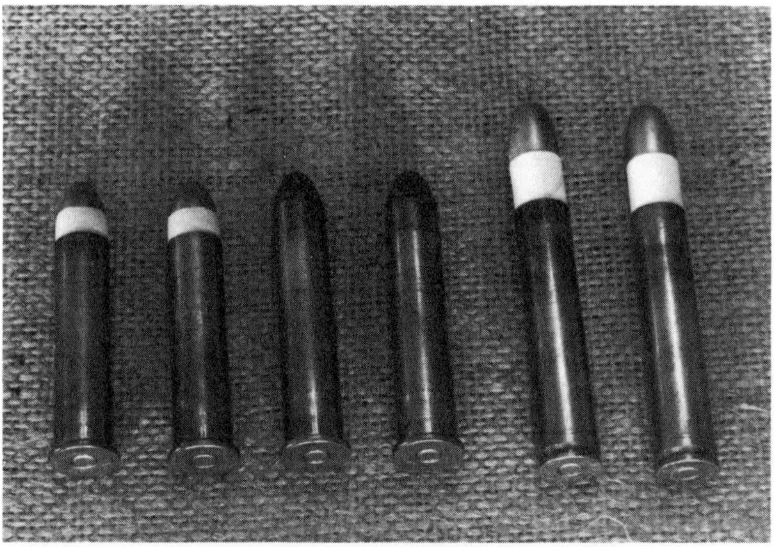

Several .45-70 cartridges and .458 Winchester cartridges loaded with paper patched bullets. Note the amount of exposed patch on these bullets. These cartridges are loaded for use in Ruger single shots having a freebored chamber or, in the case of the .458, a long tapered throat. The two dark bullets in the center have been waterproofed and rubbed with graphite.

Testing cartridges loaded with paper patched bullets in the .458. Note the wraparound grip used on the rifle's forend. When shooting rifles of heavy recoil, either off the bench or while hunting, such a grip gives far better control of the rifle.

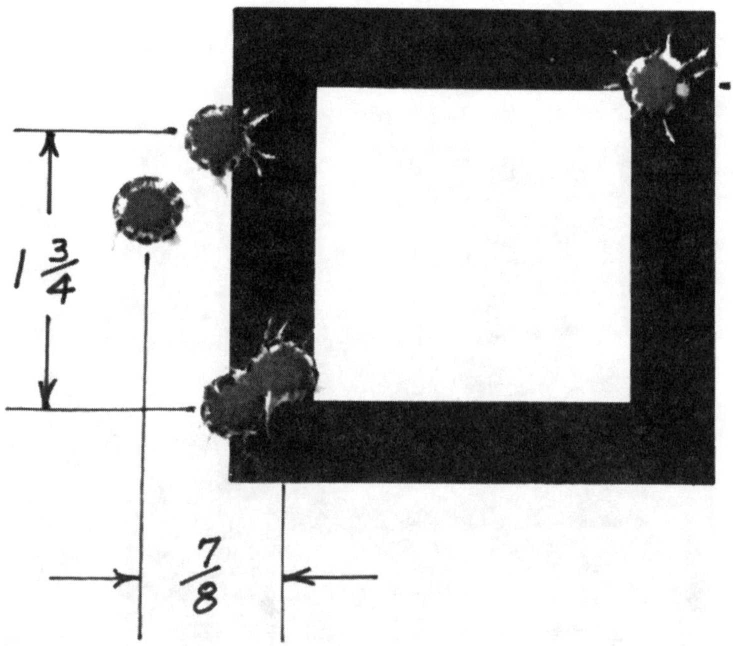

This group shot with a Marlin 1895SS .45-70 shows what can be done with soft paper patched bullets in a lever action rifle at 100 yards. The load was 50 grains of IMR-3031 behind a 400-grain waterproofed paper patched bullet.

Paper patched hunting loads for the Marlin 1895SS .45-70. These bullets are from an adjustable nose-pour Hoch mould. Note that these bullets are seated much deeper in the case than those loaded for the Ruger rifles having the freebored chambers. This is necessary for the Marlin so that they will feed through the action.

Ruger No. 3 .45-70 Kb scope
54 Gr. RS Pyrodex w/Lube Ball
419 gr. HP paper patch pure lead
100 yds. 1⅝ center-to-center

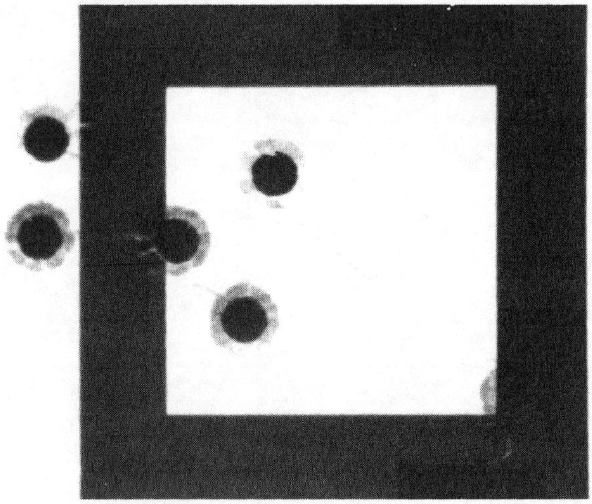

This group was made by using 419-grain, pure-lead, hollowpoint, paper patched bullets over 54 grains of RS Pyrodex and a lube wad. The bullets measured .453 inch over the patch before firing. Group measures 1⅝ inches center to center.

12

Theory of the Paper Patched Bullet

It seems like one heck of a time to start discussing the theory of paper patched bullets after having spent several thousand words on their history, use and production. Yet, at the beginning of this text the fact that paper patched bullets do work was considered far more important than why they work. After using several thousand of them in various calibers, and observing their performance, one begins to wonder what really makes the paper patched bullet so different from the grooved lubricated bullet.

Why would a few thousandths of an inch of paper enable one to shoot a pure lead bullet at 2,198 fps from a 10-inch twist .30-30 when a naked grooved bullet of the same material cannot be driven from a slow-twist barrel at much over 1,400 fps without severe leading and a dramatic loss of accuracy? Do soft, grooved bullets skid over the tops of the lands at the higher velocities? If so, why doesn't the paper patched bullet do the same?

First, we should examine the functions of the paper patch. What does it do?

Most important, as previously mentioned, the paper patch prevents a metal-to-metal contact of the bullet with the barrel. If the patch should fail in this function, the bullet upon which it fails is doomed to inaccurate flight. In fact,

I firmly believe that when using paper patched bullets, nothing except the paper patch and its attendant lubricant should ever touch the barrel's bore. I realize this is contrary to some bullet designs (bullets that seem to be satisfactory at the lower velocities). Yet, I have to believe that these same bullets would deliver better performance at all velocities and with softer alloys if there was no metal-to-metal contact with the barrel.

In addition to its function as a nonfouling bearing surface, the paper patch also seals the bore of the barrel against gas escape—a vital function regardless of the projectile type. Upon initial ignition in a cartridge rifle, there always is a certain degree of gas escape. Some barrels are freebored purposely to provide some gas escape and thus reduce pressures. Although this *might* be acceptable for metal jacketed bullets, it can raise havoc with a grooved lubricated bullet. Hot gases rushing around the base of a naked grooved bullet tend to wash away some of the metal, thus leading the first few inches of rifling.

Gas cutting, with its subsequent leading, is not a problem with paper patched bullets. The paper patch serves as an insulator to the bullet and, further, it's a good practice to patch the bullet up to the throat diameter to help prevent escaping gas and to center the back end of the bullet in the throat.

Now that we recognize the paper patch's function let's take a closer look at what happens to the bullet itself upon firing. A microsecond after ignition, rapidly expanding powder gases slam against the bullet's base with several thousand pounds of pressure per square inch. Simultaneously there are three major forces holding the bullet back against the pressure of the powder gases: the weight of the bullet or inertia, barrel friction as the rifling engages the bullet, and the column of air in the barrel ahead of the bullet which must be compressed and expelled from the barrel.

Visualizing this situation, it readily becomes apparent that the bullet is squeezed or compressed between the opposing forces so that it tries to increase in diameter, and thus exert a radial thrust against the barrel's wall to grip the rifling. *The softer the bullet is, the more tightly it will tend to grip the rifling.* It will not skid across the tops of the lands.

If this is true for the paper patched bullet, however, why isn't it true for the grooved lubricated bullet? It is. The identical action takes place with both bullets. However, where the paper patched bullet has a wrapped-on nonfouling bearing surface between the bullet and the barrel, the grooved lubricated bullet makes metal-to-metal contact with the barrel. This results in high friction and an almost instantaneous sloughing away of the bullet's surface until there is nothing left to grip the rifling.

In addition to the radial thrust that helps make the bullet grip the rifling, the *smooth-sided* paper patched bullet provides about 50 percent more support against the torsional stress of the rifling than the grooved bullet because the grooves offer no support at all. Only the grooved bullets' driving bands grip the rifling.

All of this leads to three observations about paper patched bullets.

1. Softer is better in the paper patched bullet.
2. Smooth-sided is better than grooved in the paper patched bullet.
3. Quietly lurking in the background is the belief that the larger the diameter of the unpatched bullet, the better—as long as it will still accommodate a double-wrapped patch and chamber in the rifle.

Let's take a closer look at these observations and see how they stand up in actual practice.

Observation 1: Is softer better? Perhaps—but only *inside* the barrel and within certain velocity limits *outside* the barrel. In explanation of this, it is a well-known fact that the hollowbase pure lead Minie ball fired with heavy charges of black powder is sometimes badly distorted *outside* the barrel due to high muzzle pressures. With this distortion comes a sharp decrease in accuracy. That such distortion can be reduced by a thicker skirt on the Minie ball is fine, but the potential remains. With sufficient muzzle pressure, pure-lead paper patched bullets can become distorted outside the barrel. At what point this becomes a detriment, I'm not about to say. I have seen pure-lead paper patched bullets consistently give good accuracy in .30 and .375 calibers at chronographed velocities of 2,200 fps, and sometimes higher.

Distortion of pure lead bullets due to high muzzle pressure is not the only fly in the soup. Centrifugal force of the spinning bullet can cause a bullet to virtually disintegrate, and the faster you drive the bullet, the greater is its rotational velocity and the subsequent centrifugal force. Again, this does not seem to pose a problem up to 2,200 fps.

Years ago I had copper-banded grooved bullets cast 10 to 1 lead and tin disintegrate when fired from a .30-06 loaded with 56 grains of IMR-4350. I didn't have a chronograph in those days, but the bullets in question must have been touring at least 2,700 fps. Coming from a 10-inch twist barrel, that's 3,240 revolutions per second or 194,400 revolutions per minute.

When it comes to hard paper patched bullets, the late Col. E. H. Harrison drove hard (19 BHN) .30-caliber paper patched bullets up to 3,000 fps with good accuracy. Although the practicality of such bullets in the hunting field is open to argument, Harrison's experiments opened a lot of new doors for paper patched bullets.

Now for Observation 2: Is a smooth-sided paper patched bullet better than a grooved paper patched bullet? Personally, I prefer a smooth-sided paper patched bullet if for no other reason than that it casts much better than the grooved bullet. Other than that, it looks better—more traditional. If one wants to get technical, one can say that the smooth-sided bullet cannot distort as much under pressure as the grooved bullet. It has more structural support than the grooved bullet because it lacks the grooves that would allow it to squash up like an accordion under pressure.

In actual practice, the grooved bullet works just about as well as the smooth-sided bullet. If there is a difference in performance, it usually can be attributed to a poor quality casting and/or the overall design of the bullet rather than the fact that it is grooved.

Observation 3 also has its limitations. Although it's true that the larger the unpatched bullet is, the more positively it will engage the rifling, there is a point of diminishing returns. Just as pouring 10 pounds of sugar in a five-pound bag results in an overflow, so does the use of a groove diameter bullet wrapped in a .0001 inch paper jacket. Granted, I have used such bullets with fairly respectable results, but the fact remains that when such a paper patched bullet is squeezed through a barrel, there must be considerable distortion at the base which is not conducive to good accuracy.

This distortion would be far more severe with a full-diameter smooth-sided bullet than with a corresponding grooved bullet, especially if the grooved bullet had a gas check shank. With the smooth-sided bullet, all of the excess metal would be pushed to the rear where it would bulge the base. With the grooved bullet, the distortion would be distributed in each groove and around the gas check shank rather than the base itself.

In summation, we have the following rules of thumb when dealing with paper patched bullets.

1. The softer the bullet is, the more tightly it grips the rifling, but the more vulnerable it is to distortion due to muzzle pressure and centrifugal force. Velocity limits with pure lead appear to be approximately 2,200 fps. Usually a bullet slightly hardened with tin will give better accuracy than one of pure lead.
2. For best results, nothing but the paper patch and its attendant lubricant should ever touch the barrel.

3. Grooved design or smooth-sided is a matter of personal preference.
4. Ideally, the *unpatched* bullet should be about .0005 inch larger than the rifle's bore diameter.

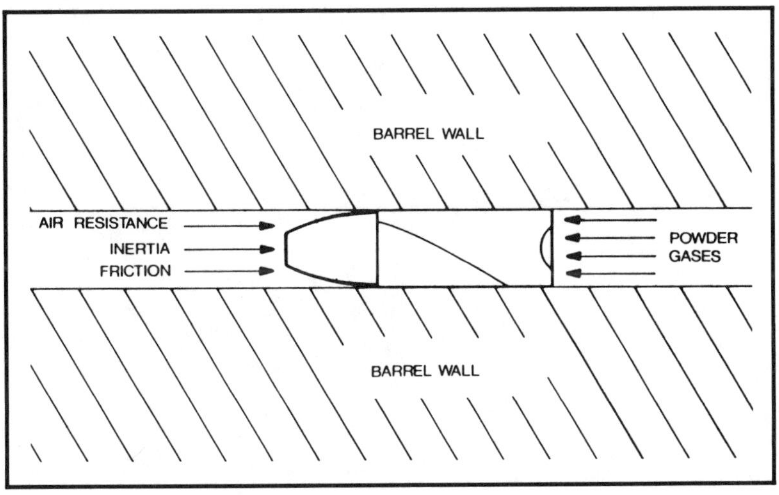

The instant a bullet is launched, the forces illustrated above go into action. During its trip down the barrel the paper patch proves its capability in preventing leading at velocities far above that of a corresponding grooved lubricated bullet.

13

Back to the Beginning

It is not likely that paper patched bullets will draw neon-light attention from those who lean heavily on modern technology. The fact remains, however, that more game has been killed on the North American continent with lead bullets and black powder than has ever been killed with jacketed bullets and smokeless powder. Further, a large portion of this game—maybe the majority of it—was killed during the years of 1870 to 1900 during the great westward expansion. This being true, it is evident that despite its primitive background and development, the paper patched bullet was a very effective projectile and played an important part in the development of the United States.

Although technology also plays a crucial part—yes, it is a vital factor—in our national progress, technology's primary function is to make things easier for mankind, hence the proliferation of screaming velocity cartridges capable of shooting a bullet from this county to the next, and powerful 3 × 9 scopes to help it get there!

How much progress is this? How much better are some of the hot-shot magnums than were the old Sharps or Remington cartridges with their paper patched bullets pushed by a small shovelful of black powder?

In Miles Gilbert's book *Getting a Stand*, there is a copy of a manuscript written by buffalo hunter Oliver Perry Hanna in the late 1920s. According to Miles Gilbert, Hanna's original manuscript is in the Wyoming State Archives in Cheyenne.

Quoting from pages 141 and 142 of Gilbert's fine book, we find these words written by Hanna:

> We arrived on the hill overlooking Young's Creek and discovered on the other side a band of about twenty blacktail. I was driving the head team and saw them first. I motioned to White and he said "We better take a shot at them before we go into camp." They were four or five hundred yards away and I was figuring which was the best way to get close to them, when I saw White preparing to shoot from that long distance. He got out one of his big sixteen pound rifles, fifty caliber, that consumed one hundred grains of powder. The first shot hit under the belly of one and the next shot killed one. Most every shot after that I could see one fall.

Lest you think this was wanton killing, Hanna and White were under contract to furnish 5,000 pounds of wild meat every week to Fort McKinney. The blacktail deer episode occurred only a day or two after their first meeting, and from that time on, Hanna left his "popgun" in the wagon and relied on one of White's three Sharps.

The blacktail deer were so far away that White shot at least nine of them before they spooked and ran. Later, after Hanna learned to use one of the big rifles, he shot elk out of a single group, and said that many times he and White got their entire quota of meat in a single day.

These men were professionals. They were in business, and then as today, time was money. They couldn't afford crippled game and they couldn't spend time crawling around on their bellies trying to get closer. Even if we cut the range that Hanna quoted all the way back to 250 yards, that would still be pretty remarkable performance for any rifle. I doubt that there are many magnum owners today who could do as well. In fact, despite the technology involved, I doubt that there are many rifle and cartridge combinations available today that would perform as well.

When one considers the type of work Hanna and White were doing and the fact that they had to depend upon consistently good bullet performance from 50 to 500 yards for shot after shot, I believe that given a choice today, they would select the soft lead paper patched bullet over any metal jacketed bullet ever produced.

In my opinion, the paper patched bullet was one of the greatest (if not the greatest) developments in firearms history.

The greatest hunting partner a man ever had! A wife who urged me to buy new rifles at a time when income was in short supply; a wife who could see a future that I dared not even dream about. Without her help and support, paper patched bullets in my life would never have happened. Here she is, my wife Betty, taking a rest at a place we call the Medicine Rocks in late March 1990. The rifle is a Marlin 1894 Sporter .44 Magnum. The load is an RCBS 44-240-SWC bullet in front of 5.0 grains of Bullseye.